Holt Mathematics

Chapter 4 Resource Book

HOLT, RINEHART AND WINSTON
A Harcourt Education Company
Orlando • Austin • New York • San Diego • London

Printed in the United States of America

ISBN 0-03-078393-3

5 170 09 08

CONTENTS

Blackline Masters

Date____________

Dear Family,

In this chapter, your child will learn about exponents, scientific notation, squares and square roots, and the real numbers.

Exponents are often used in multiplication problems when one factor is being multiplied many times.

$$4 \bullet 4 \bullet 4 \bullet 4 \bullet 4 \bullet 4 \bullet 4 = 4^7$$

The term 4^7 is called a **power.** The **exponent,** 7, represents the number of times the **base,** 4, is used as a factor.

Students learn how to multiply and divide **powers** that have the same **base.**

Multiplying Powers with the Same Base	
Rule	
To multiply powers with the same base, keep the base and add the exponents.	$3^5 \bullet 3^6 = 3^{11}$

Dividing Powers with the Same Base	
Rule	
To divide powers with the same base, keep the base and subtract the exponents.	$6^9 \div 6^4 = 6^5$

Raising a Power to a Power	
Rule	
To raise a power to a power, keep the base and multiply the exponents.	$(9^4)^5 = 9^{4(5)} = 9^{20}$

Students also learn to use **scientific notation,** which is a shorthand way to write very large numbers and very small numbers using exponents and powers of ten. A number in scientific notation is expressed as a number between 1 and 10 multiplied by a power of 10.

Standard notation: 93,000,000
Scientific notation: $9.3 \bullet 10^7$

In this example, to express 93,000,000 in scientific notation, the decimal is placed between 9 and 3 to show a number between 1 and 10. The number of decimal places to the right of 9.3 must then be calculated. Since there are 7 places to the right, the scientific notation is $9.3 \bullet 10^7$.

Standard notation: 0.00000005
Scientific notation: $5 \bullet 10^{-8}$
In the above example, in order to express 0.00000005 in scientific

notation, the decimal is moved to the right of 5 to show a number between 1 and 10. The number of decimal places to the left of 5 must then be calculated. Since there are 8 places to the left of 5, the exponent is negative 8.

To **square** a number, multiply the number by itself: $8^2 = 8 \times 8 = 64$
Taking the **square root** of a number is the inverse of squaring the number: $\sqrt{64} = 8$

Every positive number has two square roots, one positive and one negative. So, $\sqrt{64} = \pm 8$. (The symbol $\pm$ means "plus or minus" and denotes the positive and negative signs of the root.) The positive square root is called the **principal square root**.

Your child will learn to estimate square roots.

Each square root lies between two integers. Name the integers.

$\sqrt{30}$ Think: What are perfect squares close to 30?

$5^2 = 25$ $25 < 30$

$6^2 = 36$ $36 > 30$

$\sqrt{30}$ is between 5 and 6 because 30 is between 25 and 36.

Rational numbers can be written as fractions and as decimals that either terminate or repeat.

$3\frac{4}{5} = 3.8$ $\frac{2}{3} = 0.\overline{6}$ $\sqrt{1.44} = 1.2$

Irrational numbers can only be written as decimals that do *not* terminate or repeat.

Examples: $\sqrt{2}$, $\sqrt{3}$, π, 0.414213...

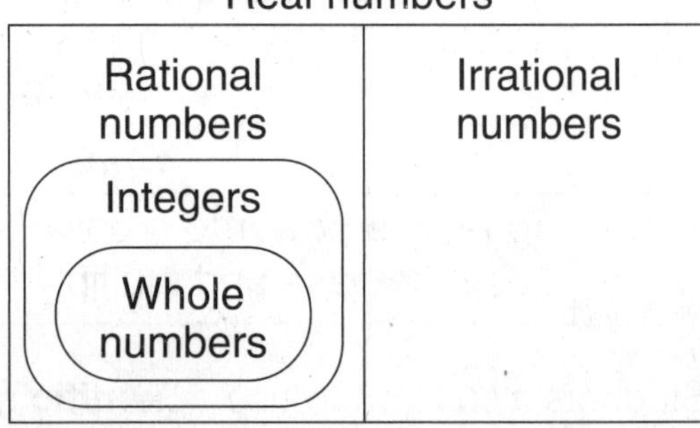

Real numbers consist of the set of rational numbers and the set of irrational numbers. Here are some examples of how we classify real numbers:

$\sqrt{3}$
irrational, real — 3 is a whole number that is not a perfect square.

-56.85
rational, real — -56.85 is a terminating decimal.

$\sqrt{\frac{9}{3}}$
whole, integer, rational, real — $\sqrt{\frac{9}{3}} = \frac{3}{3} = 1$

For additional resources, visit go.hrw.com and enter the keyword MT7 Parent.

Name ______________________ Date __________ Class __________

LESSON **4-1**

Practice A

Exponents

Name the base and exponent for each power.

1. 15^3

base ________

exponent ________

2. 8^{10}

base ________

exponent ________

3. $(-3)^2$

base ________

exponent ________

Write using exponents.

4. $2 \cdot 2 \cdot 2$

____

5. $3 \cdot 3 \cdot 3 \cdot 3 \cdot 3 \cdot 3$

____

6. $(-1) \cdot (-1) \cdot (-1) \cdot (-1)$

____

7. $(-5) \cdot (-5)$

____

8. $a \cdot a \cdot a \cdot a \cdot a$

____

9. $(-2) \cdot (-2) \cdot (-2) \cdot (-2)$

____

Evaluate.

10. 3^2 ________

11. 2^4 ________

12. 4^3 ________

13. $(-1)^4$ ________

14. $(-2)^3$ ________

15. $(-1)^5$ ________

16. 5^2 ________

17. $(-3)^4$ ________

Evaluate each expression for x = –2 and y = 3.

18. $y^3 + 1$ ________

19. $x^2 - 1$ ________

20. $x^4 + y$ ________

21. $2x - y^2$ ________

22. Find the area of square with a side of 6 m. (Hint: Area = s^2.)

____

23. Write an expression for 4 as a factor 5 times.

____

Name ____________ Date ________ Class ________

LESSON 4-1

Practice B

Exponents

Write in exponential form.

1. $6 \cdot 6 \cdot 6 \cdot 6 \cdot 6 \cdot 6$

2. $7 \cdot 7 \cdot 7 \cdot 7$

3. $(-8) \cdot (-8) \cdot (-8) \cdot (-8)$

4. $5 \cdot 5 \cdot 5 \cdot b \cdot b \cdot b \cdot b$

Evaluate.

5. 10^2

6. $(-6)^2$

7. 8^2

8. $(-7)^2$

9. $(-5)^3$

10. 12^2

11. $(-9)^2$

12. $(-4)^3$

13. 2^5

14. 5^4

15. $(-3)^4$

16. 6^3

Evaluate each expression for the given values of the variables.

17. $n^3 - 5$ for $n = 4$

18. $4x^2 + y^3$ for $x = 5$ and $y = -2$

19. $m^p + q^2$ for $m = 5$, $p = 2$, and $q = 4$

20. $a^4 + 2(b - c^2)$ for $a = 2$, $b = 4$, and $c = -1$

21. Write an expression for five times a number used as a factor three times.

22. Find the volume of a regular cube if the length of a side is 10 cm. (Hint: $V = l^3$.)

Name ______________________________ Date ____________ Class ____________

LESSON 4-1

Practice C
Exponents

Evaluate.

1. $(-10)^2$ ____________

2. 7^3 ____________

3. $(-14)^2$ ____________

4. $-(16)^2$ ____________

5. $(-2)^5$ ____________

6. $(-5)^4$ ____________

7. 3^5 ____________

8. $-(4)^4$ ____________

Evaluate each expression for the given values of the variables.

9. $n^3 - 5p$ for $n = 4$ and $p = 3$ ____________

10. $x^2 + y^3$ for $x = 5$ and $y = -3$ ____________

11. $a^b + 6^c$ for $a = 3$, $b = 4$, and $c = 2$ ____________

12. $s^4 + (t - r)^5$ for $s = 2$, $t = 7$, and $r = 8$ ____________

13. $y^x + x^y - xy$ for $x = 2$ and $y = 3$ ____________

14. $5x^9 - (y + z)$ for $x = -1$, $y = 10$, and $z = -6$ ____________

15. $x^z \div y^x$ for $x = 2$, $y = 4$, and $z = 5$ ____________

16. $ac - b^c$ for $a = 12$, $b = -6$, and $c = 2$ ____________

17. Find the area of a circle if the radius is 20 in. (Hint: $A = \pi r^2$.) Use 3.14 for π.

____

18. If the area of a regular pentagon is $A = 1.720a^2$, in which a is one of the sides. Find the area of a regular pentagon with a side that measures 50 cm.

____

Name ______________________ Date ____________ Class ____________

LESSON 4-1

Reteach

Exponents

base → 3^5 ← exponent

The fifth power of 3 $3^5 = 3 \cdot 3 \cdot 3 \cdot 3 \cdot 3$

3 used as a factor 5 times

Complete to write each expression using an exponent. State the power.

1. $5 \cdot 5 \cdot 5 \cdot 5 = 5^{\_}$

   the ____________ power of 5

2. $(-7) \cdot (-7) \cdot (-7) = (-7)^{\_}$

   the ____________ power of ____

Complete to evaluate each expression.

3. $(-2)^3 = (-2)(-2)(-2) =$ ____________

4. $10^4 =$ ____ • ____ • ____ • ____ = ____________

5. $(-5)^4 =$ (____)(____)(____)(____) = ____________

When an expression is a product that includes a power, you simplify the power first.

$3 \cdot 2^3 = 3 \cdot 2 \cdot 2 \cdot 2 = 3 \cdot 8 = 24$

Complete to simplify each expression.

6. $4 \cdot (-2)^3 = 4$(____)(____)(____) = ____________

7. $5 \cdot 3^3 =$ ____ • ____ • ____ • ____ = ____________

8. $(3 \cdot 2)^3 = 6^3 =$ ____ • ____ • ____ = ____________

9. $(-4(-2))^3 =$ (____)^3 = (____)(____)(____) = ____________

10. $25 - 3(4 \cdot 3^2)$

    = 25 − 3(4 • ______)

    = 25 − 3(______)

    = 25 − ______

    = ______

11. $-100 - 2(3 \cdot 4)^2$

    = −100 − 2(______)^2

    = −100 − 2(______)

    = −100 − ______

    = ______

12. $15 - 4(3 + 3^2)$

    = 15 − 4(3 + ______)

    = 15 − 4(______)

    = 15 − ______

    = ______

Name ______________________ Date __________ Class __________

Challenge
Check This Out

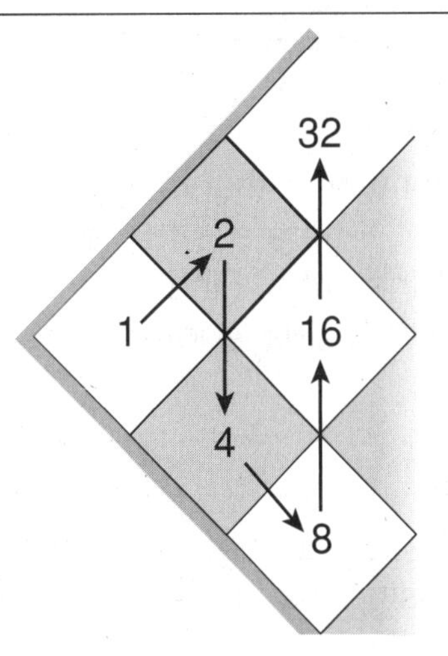

Imagine an 8 × 8 checkerboard.

- Put 1 penny on the first square.
- Stack 2 pennies on an adjacent square.
- Stack 4 pennies on an adjacent square.
- Stack 8 pennies on an adjacent square.

Assume the pattern continues so that each square has double the number of pennies as the previous square.

Complete the table.

	Square	1	2	3	4	5	6	7	8
1.	**Number of Pennies**	1	2	4	8	16			
2.	**Exponent Form**	2^0	2^1	2^2	2^3				

3. Look for a pattern in the table. How many pennies are on the 12th square? the 15th square?

____

____

4. Find the number of pennies on the 25th square. How much money is this? Estimate the height of the stack.

____

____

5. Find the number of pennies on the 10th square. Find the total number of pennies on the first nine squares. Which number is greater?

____

____

6. Explain why it is impossible to stack pennies on every square in the manner described at the top of the page.

____

____

Name ______________________________ Date ____________ Class ____________

LESSON 4-1

Problem Solving

Exponents

Write the correct answer.

1. The formula for the volume of a cube is $V = e^3$ where e is the length of a side of the cube. Find the volume of a cube with side length 6 cm.

____

2. The distance in feet traveled by a falling object is given by the formula $d = 16t^2$ where t is the time in seconds. Find the distance an object falls in 4 seconds.

____

3. The surface area of a cube can be found using the formula $S = 6e^2$ where e is the length of a side of the cube. Find the surface area of a cube with side length 6 cm.

____

4. John's father offers to pay him 1 cent for doing the dishes the first night, 2 cents for doing the dishes the second, 4 cents for the third, and so on, doubling each night. Write an expression using exponents for the amount John will get paid on the tenth night.

____

Use the table below for Exercises 5–7, which shows the number of e-mails forwarded at each level if each person continues a chain by forwarding an e-mail to 10 friends. Choose the letter for the best answer.

Forwarded E-mails

Level	E-mails forwarded
1	10
2	100
3	1000
4	10,000

5. How many e-mails were forwarded at level 5 alone?

A 5^{10}　　**C** 2^{10}

B 2^5　　**D** 10^5

6. How many e-mails were forwarded at level 6 alone?

F 100,000　　**H** 10,000,000

G 1,000,000　　**J** 100,000,000

7. Forwarding chain e-mails can create problems for e-mail servers. Find out how many total e-mails have been forwarded after 6 levels.

A 1,111,110　　**C** 1,000,000

B 6,000,000　　**D** 100,000,000

Name ______________________ Date __________ Class __________

Reading Strategies

Multiple Meanings

Exponents are an efficient way to express repeated multiplication.

Example: 4^5 → is read "4 to the fifth power."

4^5 means **4** is a factor **5** times: $4 \times 4 \times 4 \times 4 \times 4$.

$4^5 = 1024$ → is read "4 to the fifth power equals 1024," or "the value of 4 to the fifth power is 1024."

The **base** identifies the factor.
The **exponent** identifies how many times the base is a factor.

base → 4^5 ← exponent

A negative number with an even exponent will be positive. → $(-8)^4 = (-8)(-8)(-8)(-8)$

$(-8)^4 = 4096$

"negative 8 to the fourth power"

A negative number with an odd exponent will be negative. → $(-8)^3 = (-8)(-8)(-8)$

$(-8)^3 = -512$

"negative 8 to the third power"

Answer each question.

1. How do you read 7^4?

____

2. What does 7^4 mean?

____

3. What is the value of 7^4? ______________________________

4. How do you read $(-3)^5$?

____

5. What does $(-3)^5$ mean?

____

6. How you can tell when a negative number raised to a power will have a negative value?

____

7. Will the value of $(-6)^2$ be positive or negative? ______________

8. Will the value of $(-6)^5$ be positive or negative? ______________

Name ______________________ Date ____________ Class ____________

LESSON 4-1

Puzzles, Twisters & Teasers

Something's Fishy!

Rewrite each of the following using exponents. Then solve the riddle.

H 4 ______

Y $8 \cdot 8$ ______

S $(-9) \cdot (-9) \cdot (-9)$ ______

O $(-3) \cdot (-3) \cdot (-3) \cdot (-3) \cdot (-3) \cdot (-3)$ ______

I $6 \cdot 6 \cdot 6 \cdot 6 \cdot 6 \cdot 6$ ______

U $7 \cdot 7 \cdot 7 \cdot 7$ ______

F $2 \cdot 2 \cdot 2 \cdot 2 \cdot 2 \cdot 2 \cdot 2 \cdot 2 \cdot 2$ ______

C $(-6) \cdot (-6) \cdot (-6) \cdot (-6) \cdot (-6)$ ______

A $8 \cdot 8 \cdot 8 \cdot 8$ ______

N $(-12) \cdot (-12) \cdot (-12)$ ______

T $7 \cdot 7 \cdot 7$ ______

X $(-5) \cdot (-5)$ ______

What's the difference between a guitar and a fish?

___ ___ ___ ,
8^2 $(-3)^6$ 7^4

___ ___ ___ ___
$(-6)^5$ 8^4 $(-12)^3$ 7^3

___ ___ ___ ___
7^3 7^4 $(-12)^3$ 8^4

___ ___ ___ ___.
2^9 6^6 $(-9)^3$ 4^1

Name ______________________ Date __________ Class __________

LESSON 4-2

Practice A

Look for a Pattern in Integer Exponents

Evaluate the powers of 10.

1. 10^{-1} ________

2. 10^{-6} ________

3. 10^{2} ________

4. 10^{1} ________

5. 10^{0} ________

6. 10^{3} ________

7. 10^{-5} ________

8. 10^{6} ________

9. 10^{-7} ________

10. 10^{4} ________

11. 10^{-3} ________

12. 10^{5} ________

Evaluate.

13. $(-2)^{-3}$ ________

14. 3^{-4} ________

15. $(-4)^{-2}$ ________

16. 2^{-4} ________

17. 5^{-2} ________

18. 6^{-3} ________

19. $(-9)^{-2}$ ________

20. $(-3)^{-3}$ ________

21. $8 - 3^{0} + 2^{-1}$ ________

22. $4 + (-6)^{0} - 4^{-1}$ ________

23. $3(-9)^{0} + 4^{-2}$ ________

24. $6 + (-5)^{-2} - (4 + 3)^{0}$ ________

25. One centimeter equals 10^{-2} meter. Evaluate 10^{-2}.

____

26. The area of a square is 10^{-4} square feet. Evaluate 10^{-4}.

____

Name ______________________ Date __________ Class __________

LESSON 4-2

Practice B
Look for a Pattern in Integer Exponents

Evaluate the powers of 10.

1. 10^{-3} ________

2. 10^{3} ________

3. 10^{-5} ________

4. 10^{-2} ________

5. 10^{0} ________

6. 10^{4} ________

7. 10^{1} ________

8. 10^{5} ________

Evaluate.

9. $(-6)^{-2}$ ________

10. $(-9)^{-3}$ ________

11. 2^{-5} ________

12. $(-3)^{-4}$ ________

13. $(-12)^{-1}$ ________

14. 6^{-3} ________

15. $10 - (3 + 2)^{0} + 2^{-1}$ ________

16. $15 + (-6)^{0} - 3^{-2}$ ________

17. $6(8 - 2)^{0} + 4^{-2}$ ________

18. $2^{-2} + (-4)^{-1}$ ________

19. $3(1 - 4)^{-2} + 9^{-1} + 12^{0}$ ________

20. $9^{0} + 64(3 + 5)^{-2}$ ________

21. One milliliter equals 10^{-3} liter. Evaluate 10^{-3}.

____

22. The volume of a cube is 10^{6} cubic feet. Evaluate 10^{6}.

____

Name ______________________ Date __________ Class __________

LESSON 4-2

Practice C
Look for Patterns in Integer Exponents

Evaluate.

1. $(-4)^{-3}$ ________

2. 11^{-2} ________

3. 9^{-3} ________

4. 3^{-5} ________

5. $(-2)^{-6}$ ________

6. $(-4)^{-5}$ ________

7. $4^{-1} - (3 + 2)^0 + 2^{-2}$ ________

8. $7 - (-6)^0 - 3^{-2}$ ________

9. $3^2(8 + 6)^0 + 4^{-2}$ ________

10. $2^{-2} - 8 + (-4)^{-1}$ ________

11. $7(-3 - 4)^{-2} + 2^{-1}$ ________

12. $(9 - 3)^0 + 64(9 - 5)^{-2}$ ________

13. $5^2 + (-5)^0 + (4^{-3} + 2^{-6})$ ________

14. $108(4 + 2)^{-3} + (10^0 - 5)^2$ ________

15. Super Bowl XXXVI was held in New Orleans, Louisiana. The New England Patriots played the St. Louis Rams. The St. Louis fans traveled about 1127 km to New Orleans. How many meters is this? (Hint: 1 km = 10^3 m.)

____

16. The New England fans traveled 2,420,453 m. How many kilometers is this?

____

Name ______________________ Date ____________ Class ____________

LESSON **4-2**

Reteach

Look for a Pattern in Integer Exponents

To rewrite a negative exponent, move the power to the denominator of a unit fraction. $5^{-2} = \frac{1}{5^2}$

Complete to rewrite each power with a positive exponent.

1. $7^{-3} = \frac{1}{\_\_}$ **2.** $9^{-5} = \frac{1}{\_\_}$ **3.** $13^{-4} = \frac{1}{\_\_}$

Complete each pattern.

4. $10^{-1} = \frac{1}{10} = 0.1$

$10^{-2} = \frac{1}{10^2} = \frac{1}{100} = 0.01$

$10^{-3} =$ ______________________

5. $5^{-1} = \frac{1}{5}$

$5^{-2} = \frac{1}{5^2} = \frac{1}{5 \cdot 5} = \frac{1}{25}$

$5^{-3} =$ ______________________

6. $3^{-1} = \frac{1}{3}$

$3^{-2} = \frac{1}{3^2} = \frac{1}{3 \cdot 3} = \frac{1}{9}$

$3^{-3} =$ ______________________

7. $(-4)^{-1} =$ ____

$(-4)^{-2} =$ ______________________

$(-4)^{-3} =$ ______________________

Evaluate.

8. $2^{-3} = \frac{1}{\_\_} =$ __________ **9.** $(-6)^{-2} = \frac{1}{\_\_} =$ __________

10. $4^{-2} = \frac{1}{\_\_} =$ __________ **11.** $(-3)^{-3} = \frac{1}{\_\_} =$ __________

12. $6^{-2} =$ __________ **13.** $(-2)^{-3} =$ __________

14. $6^{-3} =$ __________ **15.** $(-5)^{-2} =$ __________

16. $2^{-4} =$ __________ **17.** $(-9)^{-1} =$ __________

Name ______________________ Date __________ Class __________

LESSON 4-2

Challenge

Stuff It!

$9^{\frac{1}{2}}$ means $\sqrt[2]{9^1}$.

To find the value, first evaluate the root: $\sqrt[2]{9} = 3$.

Then, raise the result to the indicated power: $3^1 = 3$.

So, $9^{\frac{1}{2}}$ is $\sqrt[2]{9^1} = 3^1 = 3$.

In general, here's the way to rewrite a term with a fractional exponent:

$$x^{\frac{a}{b}} = \sqrt[b]{x^a}$$

Evaluate $8^{\frac{2}{3}}$.

$8^{\frac{2}{3}} = \sqrt[3]{8^2}$ Rewrite using radical form.

$= 2^2$ Evaluate the root; $\sqrt[3]{8} = 2$ since $2 \cdot 2 \cdot 2 = 8$.

$= 4$ Evaluate the power.

Rewrite each term using radical form. Evaluate the root. Evaluate the power.

1. $64^{\frac{1}{2}} =$ __________
   $=$ __________
   $=$ __________

2. $100^{\frac{1}{2}} =$ __________
   $=$ __________
   $=$ __________

3. $400^{\frac{1}{2}} =$ __________
   $=$ __________
   $=$ __________

4. $64^{\frac{2}{3}} =$ __________
   $=$ __________
   $=$ __________

5. $216^{\frac{2}{3}} =$ __________
   $=$ __________
   $=$ __________

6. $1000^{\frac{2}{3}} =$ __________
   $=$ __________
   $=$ __________

7. $625^{\frac{3}{4}} =$ __________
   $=$ __________
   $=$ __________

8. $32^{\frac{2}{5}} =$ __________
   $=$ __________
   $=$ __________

9. $10{,}000^{\frac{5}{4}} =$ __________
   $=$ __________
   $=$ __________

Name ______________________________ Date ____________ Class ____________

LESSON **4-2**

Problem Solving

Look for a Pattern in Integer Exponents

Write the correct answer.

1. The weight of 10^7 dust particles is 1 gram. Evaluate 10^7.

____

2. The weight of one dust particle is 10^{-7} gram. Evaluate 10^{-7}.

____

3. As of 2001, only 10^6 rural homes in the United States had broadband Internet access. Evaluate 10^6.

____

4. Atomic clocks measure time in microseconds. A microsecond is 10^{-6} second. Evaluate 10^{-6}.

____

Choose the letter for the best answer.

5. The diameter of the nucleus of an atom is about 10^{-15} meter. Evaluate 10^{-15}.

A 0.0000000000001

B 0.00000000000001

C 0.0000000000000001

D 0.000000000000001

6. The diameter of the nucleus of an atom is 0.000001 nanometer. How many nanometers is the diameter of the nucleus of an atom?

F $(-10)^5$

G $(-10)^6$

H 10^{-6}

J 10^{-5}

7. A ruby-throated hummingbird weighs about 3^{-2} ounce. Evaluate 3^{-2}.

A -9

B -6

C $\frac{1}{9}$

D $\frac{1}{6}$

8. A ruby-throated hummingbird breathes 2×5^3 times per minute while at rest. Evaluate this amount.

F 1,000

G 250

H 125

J 30

Name ______________________ Date __________ Class __________

LESSON 4-2

Reading Strategies

Using Patterns

The pattern in this table will help you evaluate powers with exponents.

Look at the pattern of the products in the first column. You see that as you move down the column the products are getting smaller. That is because there is one less factor. Each product is divided by 2 to get the next product.

Column 1	Column 2	Column 3
$2^3 = 8$	$3^3 = 27$	$4^3 = 64$
$2^2 = 4$	$3^2 = 9$	$4^2 = 16$
$2^1 = 2$	$3^1 = 3$	$4^1 = 4$
$2^0 = 1$	$3^0 = 1$	$4^0 = 1$
$2^{-1} = \frac{1}{2}$	$3^{-1} = \frac{1}{3}$	$4^{-1} = \frac{1}{4}$
$2^{-2} = \frac{1}{4}$	$3^{-2} = \frac{1}{9}$	$4^{-2} = \frac{1}{16}$

Look at the second and third columns to answer Exercises 1-6.

1. What is the base in Column 2? ______________

2. What is the product divided by each time to get the next product? ______________

3. What is $1 \div 3$? ______________

4. What is the base in Column 3? ______________

5. What number is the product divided by each time to get the next product? ______________

6. What is $\frac{1}{4} \div 4$? ______________

Complete the table using the table above as a guide.

Column 1	Column 2	Column 3
$5^3 = 125$	$6^3 = 216$	$10^3 = 1000$
$5^2 =$	$6^2 =$	$10^2 = 100$
$= 5$	$= 6$	$= 10$

Name ______________________ Date ____________ Class ____________

LESSON **4-2**

Puzzles, Twisters & Teasers

An Alarming Activity!

Decide whether or not each equation is correct. Circle the letters above your answers. Then solve the riddle.

1. $(-2)^{-4} = 0.0002$	**O** correct	**A** incorrect
2. $10^{-5} = 0.00001$	**L** correct	**Z** incorrect
3. $10^{-2} = 1.0$	**F** correct	**A** incorrect
4. $2^{-3} = 0.3$	**D** correct	**R** incorrect
5. $10^{-3} = 0.001$	**M** correct	**S** incorrect
6. $(-4)^{-3} = 4.004$	**Q** correct	**C** incorrect
7. $10^{-4} = 0.0001$	**L** correct	**W** incorrect
8. $10^{2} = 100$	**U** correct	**E** incorrect
9. $3^{-2} = 30.0$	**T** correct	**C** incorrect
10. $10^{0} = 1$	**K** correct	**P** incorrect

What do you call a rooster that wakes you up crowing?

An ____ ____ ____ ____ ____

____ ____ ____ ____ ____

Name ______________________ Date __________ Class __________

LESSON 4-3

Practice A
Properties of Exponents

Multiply. Write the product as one power.

1. $2^2 \cdot 2^3$ ________

2. $3^5 \cdot 3^2$ ________

3. $1^3 \cdot 1^5$ ________

4. $5^4 \cdot 5^3$ ________

5. $8^1 \cdot 8^1$ ________

6. $7^4 \cdot 7^5$ ________

7. $12^1 \cdot 12^2$ ________

8. $n^3 \cdot n^8$ ________

Divide. Write the quotient as one power.

9. $\frac{2^5}{2^2}$ ________

10. $\frac{10^4}{10^3}$ ________

11. $\frac{4^6}{4^3}$ ________

12. $\frac{(-3)^6}{(-3)^4}$ ________

13. $\frac{5^8}{5^6}$ ________

14. $\frac{24^9}{24^3}$ ________

15. $\frac{(-6)^8}{(-6)^5}$ ________

16. $\frac{b^7}{b^5}$ ________

Simplify.

17. $(3^2)^4$ ________

18. $(6^3)^{-1}$ ________

19. $(4^5)^0$ ________

20. $(8^2)^3$ ________

21. $(5^{-2})^3$ ________

22. $(7^0)^4$ ________

23. $(9^4)^{-2}$ ________

24. $(s^5)^2$ ________

25. The Haywood Paper Company has 5^2 warehouses. Each warehouse holds 5^5 boxes of paper. How many boxes of paper are stored in all the warehouses? Write the answer as one power.

____

26. Write the expression for 5 used as a factor eight times being divided by 5 used as a factor six times. Simplify the expression as one power.

____

Name ______________________ Date __________ Class __________

LESSON 4-3

Practice B

Properties of Exponents

Multiply. Write the product as one power.

1. $10^5 \bullet 10^7$ **2.** $x^9 \bullet x^8$ **3.** $14^7 \bullet 14^9$ **4.** $12^6 \bullet 12^8$

5. $y^{12} \bullet y^{10}$ **6.** $15^9 \bullet 15^{14}$ **7.** $(-11)^{20} \bullet (-11)^{10}$ **8.** $(-a)^6 \bullet (-a)^7$

Divide. Write the quotient as one power.

9. $\frac{12^9}{12^2}$ **10.** $\frac{(-11)^{12}}{(-11)^8}$ **11.** $\frac{x^{10}}{x^5}$ **12.** $\frac{16^{10}}{16^2}$

13. $\frac{17^{19}}{17^2}$ **14.** $\frac{14^{15}}{14^{13}}$ **15.** $\frac{23^{17}}{23^9}$ **16.** $\frac{(-a)^{12}}{(-a)^7}$

Simplify.

17. $(6^2)^4$ **18.** $(2^4)^{-3}$ **19.** $(3^5)^{-1}$ **20.** $(y^5)^2$

21. $(9^{-2})^3$ **22.** $(10^0)^3$ **23.** $(x^4)^{-2}$ **24.** $(5^{-2})^0$

Write the product or quotient as one power.

25. $\frac{w^{12}}{w^3}$ **26.** $d^8 \bullet d^5$ **27.** $(-15)^5 \bullet (-15)^{10}$

28. Jefferson High School has a student body of 6^4 students. Each class has approximately 6^2 students. How many classes does the school have? Write the answer as one power.

29. Write the expression for a number used as a factor fifteen times being multiplied by a number used as a factor ten times. Then, write the product as one power.

Name ______________________ Date __________ Class __________

LESSON 4-3

Practice C

Properties of Exponents

Write the product or quotient as one power.

1. $x^{10} \bullet x^8$

2. $(-10)^{14} \bullet (-10)^4$

3. $\frac{d^5}{d}$

4. $9^{10} \div 9^2$

5. $t^{12} \bullet t^5$

6. $\frac{(-x)^{10}}{(-x)^8}$

7. $16^8 \div 16^8$

8. $14^9 \bullet 14^9$

9. $(-k)^{12} \div (-k)^9$

10. $15 \bullet 15^{11}$

11. $17^{10} \bullet 17$

12. $x^8 \div x^8$

Simplify.

13. $(5^3)^4$

14. $(7^4)^{-2}$

15. $(2^6)^{-1}$

16. $(x^2)^5$

17. $(3^{-2})^{-3}$

18. $(12^0)^{-2}$

19. $(w^4)^2$

20. $(y^{-1})^6$

Write as one power.

21. $x^3 \bullet x^2 \bullet x^4$

22. $5^5 \div 5^2 \bullet 5^3$

23. $8^4 \bullet 8 \div 8^2$

24. $x^7 \div x^2 \div x^3$

25. $(-4)^4 \bullet (-4)^6 \div (-4)^2$

26. $2^5 \div 2 \bullet 2^2 \div 2^3$

27. Justine and ninety-nine of her co-workers won the lottery worth $\$10^8$. They all received their winnings over ten years. How much did each receive in a one-year period?

____

28. A number to the 7th power divided by the same number to the 3rd power equals 256. What is the number?

____

Name ______________________ Date __________ Class __________

LESSON 4-3

Reteach

Properties of Exponents

To multiply powers with the same base, keep the base and add exponents.

$x^a \bullet x^b = x^{a+b}$

$4^5 \bullet 4^2 = 4^{5+2} = 4^7$

$8^3 \bullet 8 = 8^{3+1} = 8^4$

To divide powers with the same base, keep the base and subtract exponents.

$x^a \div x^b = x^{a-b}$

$4^5 \div 4^2 = 4^{5-2} = 4^3$

$8^3 \div 8 = 8^{3-1} = 8^2$

To raise a power to a power, keep the base and multiply exponents.

$(x^a)^b = x^{ab}$

$(4^5)^2 = 4^{5(2)} = 4^{10}$

Complete to see why the rules for exponents work.

1. $4^5 \bullet 4^2 = (\_\_)(\_\_)(\_\_)(\_\_)(\_\_) \bullet (\_\_)(\_\_) = 4^{\_\_}$

2. $8^3 \bullet 8 = (\_\_)(\_\_)(\_\_) \bullet (\_\_) = 8^{\_\_}$

3. $4^5 \div 4^2 = \dfrac{4^5}{4^2} = \dfrac{\cancel{4} \bullet \cancel{4} \bullet 4 \bullet 4 \bullet 4}{\cancel{4} \bullet \cancel{4}} = 4^{\_\_}$

4. $8^3 \div 8 = \dfrac{8^3}{8} = \dfrac{\cancel{8} \bullet 8 \bullet 8}{\cancel{8}} = 8^{\_\_}$

5. $(4^2)^3 = 4^2 \bullet 4^2 \bullet 4^2 = 4^{2+2+2} = 4^{2(3)} = 4^{\_\_\_}$

Complete to write each product or quotient as one power.

6. $12^3 \bullet 12^2 = 12^{3+2} = 12^{\_\_}$

7. $9^4 \bullet 9^3 = 9^{\_\_\_} = 9^{\_\_}$

8. $\dfrac{7^6}{7^2} = 7^{6-2} = 7^{\_\_}$

9. $\dfrac{12^6}{12^4} = 12^{\_\_\_} = 12^{\_\_}$

Write each product or quotient as one power.

10. $10^4 \bullet 10^6 =$ ________

11. $5^5 \bullet 5 =$ ________

12. $4^5 \bullet 4 \bullet 4^3 =$ ________

13. $\dfrac{15^6}{15^2} =$ ________

14. $\dfrac{9^5}{9} =$ ________

15. $\dfrac{2^{10}}{2^2} =$ ________

Simplify.

16. $(5^3)^4 = 5^{3(4)} =$ ________

17. $(6^2)^4 = 6^{2(4)} =$ ________

18. $(2^5)^2 =$ ________

Name ______________________ Date __________ Class __________

LESSON 4-3

Challenge

Square Dance

Study these patterns.

$1 = 1^2$

$1^2 + 1 + 2 = 4 = 2^2$

$2^2 + 2 + 3 = 9 = 3^2$

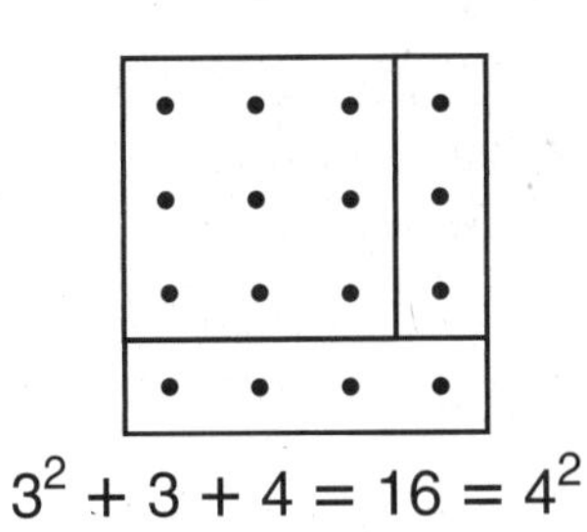

$3^2 + 3 + 4 = 16 = 4^2$

So, according to the pattern, 5^2 can be written as the sum of 4^2 and two consecutive integers.

1. Draw a diagram and write an equation to illustrate 5^2.

Equation: ______________________

2. Draw a diagram and write an equation to illustrate 8^2.

Equation: ______________________

3. Use the pattern to write an equation to indicate that, for any integer n, $(n + 1)^2$ can be written as the sum of n^2 and two consecutive integers.

Equation: ______________________

4. If you know that $20^2 = 400$, use the pattern to calculate 21^2.

$21^2 =$ ______________________

Name ______________________ Date __________ Class __________

LESSON 4-3

Problem Solving

Properties of Exponents

Write each answer as a power.

1. Cindy separated her fruit flies into equal groups. She estimates that there are 2^{10} fruit flies in each of 2^2 jars. How many fruit flies does Cindy have in all?

____

2. Suppose a researcher tests a new method of pasteurization on a strain of bacteria in his laboratory. If the bacteria are killed at a rate of 8^9 per sec, how many bacteria would be killed after 8^2 sec?

____

3. A satellite orbits the earth at about 13^4 km per hour. How long would it take to complete 24 orbits, which is a distance of about 13^5 km?

____

4. The side of a cube is 3^4 centimeters long. What is the volume of the cube? (Hint: $V = s^3$.)

____

Use the table to answer Exercises 5–6. The table describes the number of people involved at each level of a pyramid scheme. In a pyramid scheme each individual recruits so many others to participate who in turn recruit others, and so on. Choose the letter of the best answer.

Pyramid Scheme

Each person recruits 5 others.

Level	Total Number of People
1	5
2	5^2
3	5^3
4	5^4

5. Using exponents, how many people will be involved at level 6?

A 6^6 **C** 5^5

B 6^5 **D** 5^6

6. How many times more people will be involved at level 6 than at level 2?

F 5^4 **H** 5^5

G 5^3 **J** 5^6

7. There are 10^3 ways to make a 3-digit combination, but there are 10^6 ways to make a 6-digit combination. How many times more ways are there to make a 6-digit combination than a 3-digit combination?

A 5^{10} **C** 2^5

B 2^{10} **D** 10^3

8. After 3 hours, a bacteria colony has $(25^3)^3$ bacteria present. How many bacteria are in the colony?

F 25^1 **H** 25^9

G 25^6 **J** 25^{33}

Name ______________________ Date ___________ Class ___________

LESSON **4-3**

Reading Strategies

Organization Patterns

There are some rules that make multiplying or dividing exponents with the same base easier.

To multiply powers with the same base, add exponents.

$$(4 \bullet 4) \bullet (4 \bullet 4 \bullet 4) = 4 \bullet 4 \bullet 4 \bullet 4 \bullet 4$$

$$\downarrow \qquad \downarrow \qquad \downarrow$$

$$4^2 \bullet 4^3 = 4^5$$

The base of 4 is the same, so: $4^2 \bullet 4^3 = 4^{2+3} = 4^5$.

To divide powers with the same base, subtract the exponents.

$$\frac{6 \bullet 6 \bullet 6 \bullet 6 \bullet 6 \bullet 6}{6 \bullet 6 \bullet 6} = \frac{6^6}{6^3}$$

$$\downarrow \qquad \downarrow$$

$$\frac{6 \bullet 6 \bullet 6 \bullet \cancel{6} \bullet \cancel{6} \bullet \cancel{6}}{\cancel{6} \bullet \cancel{6} \bullet \cancel{6}} = 6^3$$

The base of 6 is the same, so: $\frac{6^6}{6^3} = 6^{6-3} = 6^3$

Answer each question.

1. What is the base for 3^2? ______________________
2. What is the base for 3^4? ______________________
3. Are the bases the same for these powers? ______________________
4. Write all the factors for $3^2 \bullet 3^4$. ______________________
5. Add the exponents for 3^2 and 3^4 and rewrite the number using the same base. ______________________
6. Are the answers for Exercise 4 and Exercise 5 the same? ______________________
7. Are the bases the same for $5^5 \div 5^2$? ______________________
8. Subtract the exponents and rewrite the problem. ______________________

Name ______________________ Date __________ Class __________

LESSON 4-3

Puzzles, Twisters & Teasers

Button It Up!

Find and circle words from the list in the word search. Then find a word that answers the riddle. Circle it and write it on the line.

factor	power	property	exponent	group
base	combine	evaluate	zero	apply

E	G	R	O	U	P	O	W	E	R	A	C
V	X	E	R	T	Y	Z	F	G	H	J	O
A	Y	P	R	O	P	E	R	T	Y	A	M
L	W	E	O	C	V	R	V	B	N	P	B
U	U	I	O	N	P	O	O	L	P	P	I
A	D	F	G	B	E	L	L	Y	I	L	N
T	B	A	S	E	S	N	M	J	A	Y	E
E	P	L	M	F	A	C	T	O	R	Q	E

What kind of button can't you unbutton?

a ___ ___ ___ ___ ___ button

Name ______________________ Date ___________ Class ___________

LESSON **4-4**

Practice A

Scientific Notation

Write each number in standard notation.

1. 1.76×10^{-1} ________

2. 8.9×10^{3} ________

3. 6.2×10^{-2} ________

4. 1.01×10^{2} ________

5. 5.8×10^{4} ________

6. 8.1×10^{5} ________

7. 3.8×10^{-4} ________

8. 2.03×10^{-3} ________

9. 5.0×10^{3} ________

10. 3.12×10^{5} ________

11. 7.6×10^{-2} ________

12. 8.54×10^{-5} ________

Write each number in scientific notation.

13. 376,000 ________

14. 9,580,000 ________

15. 650 ________

16. 1006 ________

17. 29 ________

18. 0.0061 ________

19. 0.0107 ________

20. 0.0002008 ________

21. 0.00053 ________

22. 250,800 ________

23. 0.000094 ________

24. 0.00086 ________

25. Earth is about 93,000,000 miles from the Sun. Write this number in scientific notation.

____

26. The diameter of Earth is about 1.276×10^{4} kilometers. The diameter of Venus is about 1.21×10^{4} kilometers. Which planet has the greater diameter, Earth or Venus?

____

Name ______________________ Date __________ Class __________

LESSON 4-4

Practice B

Scientific Notation

Write each number in standard notation.

1. 2.54×10^{2} **2.** 6.7×10^{-2} **3.** 1.14×10^{3} **4.** 3.8×10^{-1}

________ ________ ________ ________

5. 7.53×10^{-3} **6.** 5.6×10^{4} **7.** 9.1×10^{5} **8.** 6.08×10^{-4}

________ ________ ________ ________

9. 8.59×10^{5} **10.** 3.331×10^{6} **11.** 7.21×10^{-3} **12.** 5.88×10^{-4}

________ ________ ________ ________

Write each number in scientific notation.

13. 75,000,000 **14.** 208 **15.** 907,100

________ ________ ________

16. 56 **17.** 0.093 **18.** 0.00006

________ ________ ________

19. 0.00852 **20.** 0.0505 **21.** 0.003007

________ ________ ________

22. 5226 **23.** 0.04 **24.** 98,856

________ ________ ________

25. Jupiter is about 778,120,000 kilometers from the Sun. Write this number in scientific notation.

____

26. The *E. coli* bacterium is about 5×10^{-7} meters wide. A hair is about 1.7×10^{-5} meters wide. Which is wider, the bacterium or the hair?

____

Name ______________________ Date __________ Class __________

LESSON 4-4

Practice C

Scientific Notation

Write each number in standard notation.

1. 6.34×10^5 ____________

2. 7.1×10^{-5} ____________

3. 4.23×10^6 ____________

4. 8.235×10^{-7} ____________

5. 6.0089×10^{-8} ____________

6. 5.2×10^9 ____________

7. 2.0547×10^9 ____________

8. 8.394×10^{-7} ____________

9. 9.688×10^{11} ____________

Write each number in scientific notation.

10. 97,406,000,000 ____________

11. 0.000000000067 ____________

12. 5,280,000,000 ____________

13. 0.000000084 ____________

14. 652,100,000,000 ____________

15. 0.0000000007254 ____________

16. 20,509,000,000,000 ____________

17. 0.000000000000066 ____________

18. 30,500,000,000,000 ____________

19. The volume of the earth is approximately 260,000,000,000 mi^3. Write this number in scientific notation.

____

20. If light travels 10,000 km in 3.3×10^{-2} sec, how long will it take light to travel one meter?

____

21. The diameter of a human hair can measure from 1.7×10^{-5} meter to 1.8×10^{-4} meter. Which is the greater diameter?

____

Name ______________________ Date __________ Class __________

LESSON 4-4

Reteach

Scientific Notation

Standard Notation	Scientific Notation	
	(1st factor is between 1 and 10.)	(2nd factor is an integer power of 10.)
430,000	4.3×10^5	positive integer for large number
0.0000057	5.7×10^{-6}	negative integer for small number

To convert from scientific notation, look at the power of 10 to tell how many places and which way to move the decimal point.

Complete to write each in standard notation.

	1. 4.12×10^6	**2.** 3.4×10^{-5}
Is the exponent positive or negative?	________	________
Move the decimal point right or left? How many places?	________	________
Write the number in standard notation.	________	________

Write each number in standard notation.

3. 8×10^5 ________

4. 7.1×10^{-4} ________

5. 3.14×10^8 ________

To convert to scientific notation, determine the factor between 1 and 10. Then determine the power of 10 by counting from the decimal point in the first factor to the decimal point in the given number.

Complete to write each in scientific notation.

	6. 32,000,000	**7.** 0.0000000712
What is the first factor?	________	________
From its location in the first factor, which way must the decimal move to its location in the given number? How many places?	________	________
Write the number in scientific notation.	________	________

Write each number in scientific notation.

8. 41,000,000 ________

9. 0.0000000643 ________

10. 1,370,000,000 ________

Name ________________________ Date __________ Class __________

LESSON 4-4

Challenge

The Wild Blue Yonder

Astronomers measure distances within our solar system in *astronomical units* (AU).

1 AU ≈ 92,956,000 mi or 149,600,000 km
(the distance from Earth to the Sun)

Mean Distance From the Sun

Planet	km	Scientific Notation	AU
Mercury	57,900,000		0.4
Venus	108,200,000		
Earth	149,600,000		1.0
Mars	227,900,000		
Jupiter	778,400,000		
Saturn	1,429,400,000		
Uranus	2,875,000,000		
Neptune	4,504,300,000		
Pluto*	5,900,100,000		

* designated as a dwarf planet in 2006

1. The table gives each planet's mean distance from the Sun in kilometers. Write these distances in scientific notation.

2. Convert to AUs by dividing each planet's mean distance from the Sun by 1.496×10^8. Use scientific notation. Round your answers to the nearest tenth of an AU.

Example $\frac{5.79 \times 10^7}{1.496 \times 10^8} = \frac{5.79}{1.496} \times \frac{10^7}{10^8} = 3.87 \times 10^{-1} = 0.387 \approx 0.4$ AU

3. Approximately how many times greater is Saturn's distance from the Sun than is Earth's? Answer to the nearest tenth.

____

4. Approximately how many times greater is the distance of the farthest planet from the Sun than is the distance of the closest planet to the Sun? Answer to the nearest tenth.

____

Name ______________________________ Date ____________ Class ____________

LESSON 4-4

Problem Solving

Scientific Notation

Write the correct answer.

1. In June 2001, the Intel Corporation announced that they could produce a silicon transistor that could switch on and off 1.5 trillion times a second. Express the speed of the transistor in scientific notation.

____

2. With this transistor, computers will be able to do 1×10^9 calculations in the time it takes to blink your eye. Express the number of calculations using standard notation.

____

3. The elements in this fast transistor are 20 nanometers long. A nanometer is one-billionth of a meter. Express the length of an element in the transistor in meters using scientific notation.

____

4. The length of the elements in the transistor can also be compared to the width of a human hair. The length of an element is 2×10^{-3} times smaller than the width of a human hair. Express 2×10^{-3} in standard notation.

____

Use the table to answer Exercises 5–9. Choose the best answer.

Distance From Earth To Stars
Light-Year = 5,880,000,000,000 mi.

Star	Constellation	Distance (light-years)
Sirius	Canis Major	8
Canopus	Carina	650
Alpha Centauri	Centaurus	4
Vega	Lyra	23

5. Express a light-year in miles using scientific notation.

   A 58.8×10^{11} **C** 588×10^{10}

   B 5.88×10^{12} **D** 5.88×10^{-13}

6. How many miles is it from Earth to the star Sirius?

   F 4.705×10^{12} **H** 7.35×10^{12}

   G 4.704×10^{13} **J** 7.35×10^{11}

7. How many miles is it from Earth to the star Canopus?

   A 3.822×10^{15} **C** 3.822×10^{14}

   B 1.230×10^{15} **D** 1.230×10^{14}

8. How many miles is it from Earth to the star Alpha Centauri?

   F 2.352×10^{13} **H** 2.352×10^{14}

   G 5.92×10^{13} **J** 5.92×10^{14}

9. How many miles is it from Earth to the star Vega?

   A 6.11×10^{13} **C** 6.11×10^{14}

   B 1.3524×10^{13} **D** 1.3524×10^{14}

Name ______________________ Date __________ Class __________

LESSON 4-4 Reading Strategies

Organization Patterns

You can use **powers of 10** to write very large or very small numbers in a shortened form. This efficient method is called **scientific notation.** It is also useful in performing multiplication and division of very large and very small numbers.

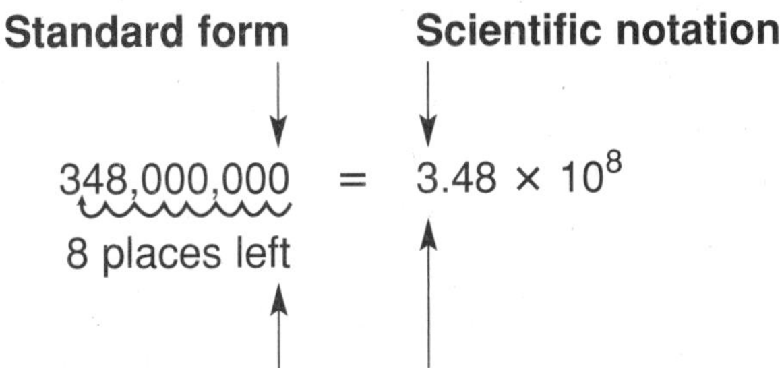

Move the decimal point to create a number between 1 and 10.

The number of places the decimal point is moved to the left is the positive exponent.

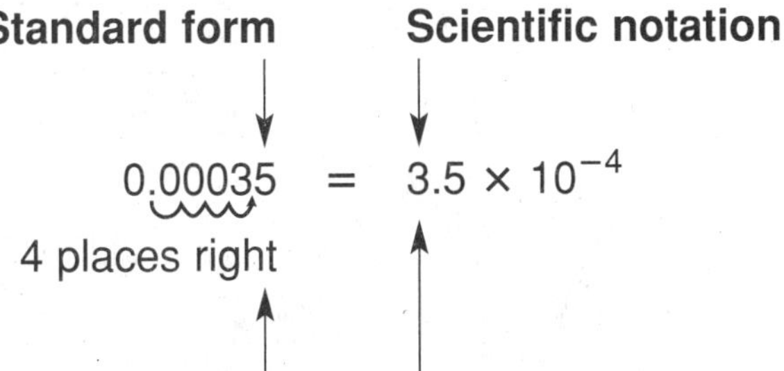

Move the decimal point to create a number between 1 and 10.

The number of places the decimal point is moved to the right is the negative exponent.

Use 0.000078 to answer Exercises 1–4.

1. How many places must you move the decimal point to create a number between 1 and 10? ______________________

2. Which direction will you move the decimal point?

   ____

3. Will the exponent be negative or positive? ______________________

4. Write the number in scientific notation. ______________________

Use 312,000,000 to answer Exercises 5–7.

5. How many places must you move the decimal point to create a number between 1 and 10? ______________________

6. Which direction will you move the decimal point? ______________________

Name ______________________ Date ____________ Class ____________

LESSON 4-4

Puzzles, Twisters & Teasers

Be a Math Hog!

Write the correct exponents to show each number in scientific notation. Then solve the riddle.

I 580,000 = 5.8 × $10^{\_\_}$

E 26,400,000 = 2.64 × $10^{\_\_}$

N 0.000135 = 1.35 × $10^{\_\_}$

A 0.000002 = 2 × $10^{\_\_}$

C 155,000,000 = 1.55 × $10^{\_\_}$

H 0.00000014 = 1.4 × $10^{\_\_}$

N 0.0003 = 3 × $10^{\_\_}$

M 7,800,000 = 7.8 × $10^{\_\_}$

L 0.0467 = 4.67 × $10^{\_\_}$

B 3500 = 3.5 × $10^{\_\_}$

U 900 = 9 × $10^{\_\_}$

X 1,000,000,000 = 1 × $10^{\_\_}$

How did the pig get to the hospital?

___ ___   ___
5   −4   −6

___ ___ ___ ___ ___ ___ ___ ___ ___ ___
−7   −6   6   3   2   −2   −6   −4   8   7

Name ________________________ Date __________ Class __________

LESSON **4-5**

Practice A

Squares and Square Roots

Find the two square roots of each number.

1. 16 ________

2. 49 ________

3. 1 ________

4. 25 ________

5. 100 ________

6. 4 ________

7. 81 ________

8. 64 ________

Evaluate each expression.

9. $\sqrt{8 + 1}$ ________

10. $\sqrt{7 - 6}$ ________

11. $\sqrt{18 - 2}$ ________

12. $\sqrt{31 + 5}$ ________

13. $\sqrt{36} + 10$ ________

14. $15 - \sqrt{25}$ ________

15. $\sqrt{49} - \sqrt{4}$ ________

16. $\sqrt{16} + 9$ ________

17. $\sqrt{\frac{64}{16}}$ ________

18. $5\sqrt{9}$ ________

19. $\sqrt{\frac{100}{4}}$ ________

20. $-3\sqrt{81}$ ________

Switzerland's flag is a square, unlike other flags that are rectangular.

21. If the flag of Switzerland has an area of 81 ft^2, what is the length of each of its sides? (Hint: $s = \sqrt{A}$)

____

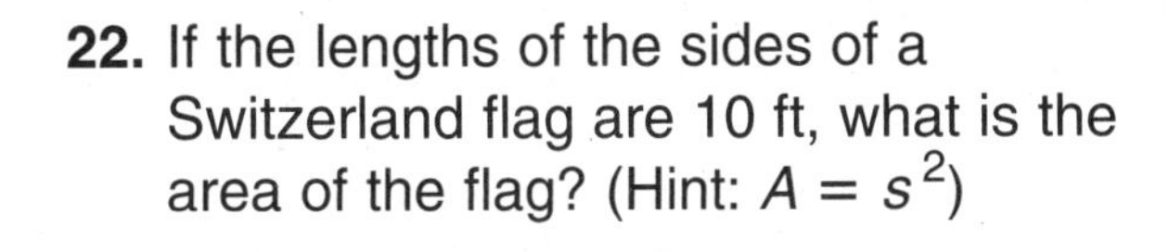

22. If the lengths of the sides of a Switzerland flag are 10 ft, what is the area of the flag? (Hint: $A = s^2$)

____

Name ______________________ Date __________ Class __________

LESSON 4-5

Practice B

Squares and Square Roots

Find the two square roots of each number.

1. 36 ________

2. 81 ________

3. 49 ________

4. 100 ________

5. 64 ________

6. 121 ________

7. 25 ________

8. 144 ________

Evaluate each expression.

9. $\sqrt{32 + 17}$ ________

10. $\sqrt{100 - 19}$ ________

11. $\sqrt{64 + 36}$ ________

12. $\sqrt{73 - 48}$ ________

13. $2\sqrt{64} + 10$ ________

14. $36 - \sqrt{36}$ ________

15. $\sqrt{100} - \sqrt{25}$ ________

16. $\sqrt{121} + 16$ ________

17. $\sqrt{\frac{25}{4}} + \frac{1}{2}$ ________

18. $\sqrt{\frac{100}{25}}$ ________

19. $\sqrt{\frac{196}{49}}$ ________

20. $3(\sqrt{144} - 6)$ ________

The Pyramids of Egypt are often called the first wonder of the world. This group of pyramids consists of Menkaura, Khufu, and Khafra. The largest of these is Khufu, sometimes called Cheops. During this time in history, each monarch had his own pyramid built to bury his mummified body. Cheops was a king of Egypt in the early 26th century B.C. His pyramid's original height is estimated to have been 482 ft. It is now approximately 450 ft. The estimated completion date of this structure was 2660 B.C.

21. If the area of the base of Cheops' pyramid is 570,025 ft^2, what is the length of one of the sides of the ancient structure? (Hint: $s = \sqrt{A}$)

____

22. If a replica of the pyramid were built with a base area of 625 in^2, what would be the length of each side? (Hint: $s = \sqrt{A}$)

____

Name ______________________ Date __________ Class __________

LESSON 4-5

Practice C

Squares and Square Roots

Find the two square roots of each number.

1. 225 ________

2. 576 ________

3. 361 ________

4. 625 ________

5. 400 ________

6. 729 ________

7. 1024 ________

8. 2500 ________

Evaluate each expression.

9. $\sqrt{\frac{1}{9}}$ ________

10. $\sqrt{\frac{144}{4}}$ ________

11. $\sqrt{\frac{64}{16}} + 3$ ________

12. $\sqrt{\frac{100}{4}} + \sqrt{4}$ ________

13. $\sqrt{\frac{81}{324}}$ ________

14. $\sqrt{196} - \sqrt{49}$ ________

15. $6(\sqrt{225} - 9)$ ________

16. $-(\sqrt{144}\sqrt{64})$ ________

Evaluate each expression.

17. $-2\sqrt{\frac{144}{36}} - 4$ ________

18. $\sqrt{676} - 15$ ________

19. $49 - \sqrt{441} + \sqrt{169}$ ________

20. The distance from base to base, home plate included, on a baseball field is 90 ft. The bases form a square. What is its area?

____

Switzerland's flag is a square, unlike other flags that are rectangular.

21. Suppose a seamstress is making a flag that has to have an area less than 50 square meters. What is the longest length, to the nearest tenth, the sides can be?
(Hint: $s = \sqrt{A}$)

____

Name ______________________ Date __________ Class __________

LESSON 4-5

Reteach

Squares and Square Roots

A **perfect square** has two identical factors.

$25 = 5 \times 5 = 5^2$ **or** $25 = (-5) \times (-5) = (-5)^2$ then 25 is a perfect square.

**Tell if the number is a perfect square.
If yes, write its identical factors.**

1. 121 ______________ **2.** 200 ______________

3. 400 ______________

Since $5^2 = 25$ and also $(-5)^2 = 25$, both 5 and −5 are **square roots** of 25.

$\sqrt{25} = 5$ and $-\sqrt{25} = -5$

The **principal square root** of 25 is 5: $\sqrt{25} = 5$

Write the two square roots of each number.

4. $\sqrt{81} =$ ________ **5.** $\sqrt{625} =$ ________ **6.** $\sqrt{169} =$ ________

$-\sqrt{81} =$ ________ $-\sqrt{625} =$ ________ $-\sqrt{169} =$ ________

Write the principal square root of each number.

7. $\sqrt{144} =$ ________ **8.** $\sqrt{6400} =$ ________ **9.** $\sqrt{10{,}000} =$ ________

Use the principal square root when evaluating an expression. For the order of operations, do square root first, as you would an exponent.

$5\sqrt{100} - 3$
$5(10) - 3$
$50 - 3$
47

Complete to evaluate each expression.

10. $3\sqrt{144} - 20$

$3 \times$ ______ $- 20$

______ $- 20$

______

11. $\sqrt{25 + 144} + 13$

______ $+ 13$

______ $+ 13$

______

12. $\sqrt{\frac{100}{25}} + \frac{1}{2}$

$\frac{\sqrt{100}}{\sqrt{25}} + \frac{1}{2}$

$\frac{\quad}{5} + \frac{1}{2}$

______ $+ \frac{1}{2}$

______

Name ______________________ Date ____________ Class ____________

LESSON 4-5

Challenge

Dig It!

Find the **digital root** of a number by adding its digits, adding the digits of the result, and so on, until the result is a single digit.

358 ➤ 3 + 5 + 8 = 16 ➤ 1 + 6 = 7 The digital root of 358 is 7.

1. Complete the table to find the digital roots of the squares of 1–17.

		Digital Root		
Number	**Square**	**Calculation**		
1			=	
2			=	
3			=	
4			=	
5			=	
6			=	
7	49	4 + 9 = 13 ➤ 1 + 3	=	4
8			=	
9			=	
10			=	
11			=	
12			=	
13			=	
14			=	
15			=	
16			=	
17			=	

2. Make an observation about the results.

____

3. Make a conjecture about the digital root of any whole-number perfect square. Verify your conjecture by using at least three more perfect squares.

____

4. A **palindrome** is a number that is the same when read forward or backward, such as 14741. Find two palindromes in the table.

____

Name ______________________ Date __________ Class __________

LESSON **4-5**

Problem Solving

Squares and Square Roots

Write the correct answer.

1. For college wrestling competitions, the NCAA requires that the wrestling mat be a square with an area of 1764 square feet. What is the length of each side of the wrestling mat?

   ____

2. For high school wrestling competitions, the wrestling mat must be a square with an area of 1444 square feet. What is the length of each side of the wrestling mat?

   ____

3. The Japanese art of origami requires folding square pieces of paper. Elena begins with a large sheet of square paper that is 169 square inches. How many squares can she cut out of the paper that are 4 inches on each side?

   ____

4. When the James family moved into a new house they had a square area rug that was 132 square feet. In their new house, there are three bedrooms. Bedroom one is 11 feet by 11 feet. Bedroom two is 10 feet by 12 feet and bedroom three is 13 feet by 13 feet. In which bedroom will the rug fit?

   ____

Choose the letter for the best answer.

5. A square picture frame measures 36 inches on each side. The actual wood trim is 2 inches wide. The photograph in the frame is surrounded by a bronze mat that measures 5 inches. What is the maximum area of the photograph?

   A 841 sq. inches **B** 900 sq. inches
   C 1156 sq. inches **D** 484 sq. inches

6. To create a square patchwork quilt wall hanging, square pieces of material are sewn together to form a larger square. Which number of smaller squares can be used to create a square patchwork quilt wall hanging?

   F 35 squares **G** 64 squares
   H 84 squares **J** 125 squares

7. A can of paint claims that one can will cover 400 square feet. If you painted a square with the can of paint, how long would it be on each side?

   A 200 feet **B** 65 feet
   C 25 feet **D** 20 feet

8. A box of tile contains 12 square tiles. If you tile the largest possible square area using whole tiles, how many tiles will you have left from the box?

   F 9 **G** 6
   H 3 **J** 0

Name ______________________ Date __________ Class __________

LESSON **4-5**

Reading Strategies

Connect Words with Symbols

A **square root** produces a given number when multiplied by itself. The large square shown below is 4 squares long on each side and has 16 squares. 4 times 4 equals 16. 4 is the **square root** of 16.

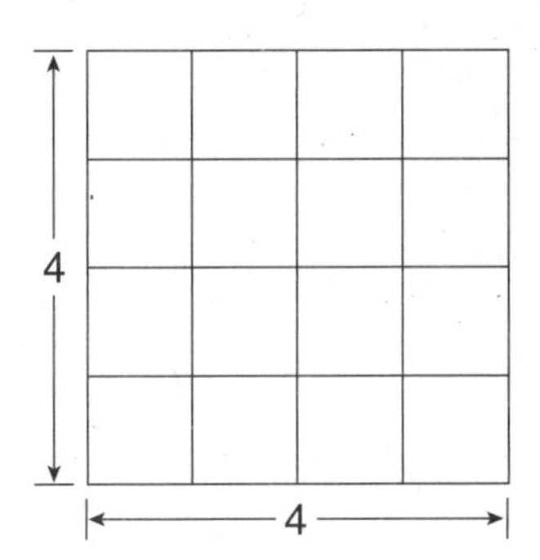

The 4 × 4 square can be described with symbols and with words.

Symbols	**Symbols**		**Words**
$4 \cdot 4 = 16$	$4^2 = 16$	→	Four squared equals sixteen.

This sign represents square root: $\sqrt{}$

$\sqrt{16} = 4$ → Read "The square root of 16 equals 4."

$\sqrt{25} = 5$ → Read "The square root of 25 equals 5."

Compare the symbols for "squared" and "square root."

$4^2 = 16$ and $\sqrt{16} = 4$

$5^2 = 25$ and $\sqrt{25} = 5$

Write in words.

1. 6^2 ______________________

2. $\sqrt{36}$ ______________________

Answer each question.

3. What is the square root of 36? ______________

4. What is the square root of 100? ______________

5. What is 7^2? ______________

Name ______________________ Date ___________ Class ___________

LESSON 4-5

Puzzles, Twisters & Teasers

Squaresville, Man!

Circle words from the list in the word search. Then find a word that answers the riddle. Circle it and write it on the line.

principal	square	root	perfect	negative
positive	solution	calculator	inverse	operation

N	M	J	P	R	I	N	C	I	P	A	L	N
E	G	H	J	O	G	B	R	L	E	J	P	O
G	M	N	B	O	S	D	U	S	R	L	O	I
A	L	K	J	T	A	X	M	E	F	O	S	T
T	C	V	B	N	M	J	B	D	E	M	I	A
I	N	V	E	R	S	E	Y	V	C	Z	T	R
V	S	Q	U	A	R	E	L	P	T	C	I	E
E	C	A	L	C	U	L	A	T	O	R	V	P
S	O	L	U	T	I	O	N	U	I	O	E	O

Why did the cookie go to see the doctor?

He was feeling ___ ___ ___ ___ ___ ___.

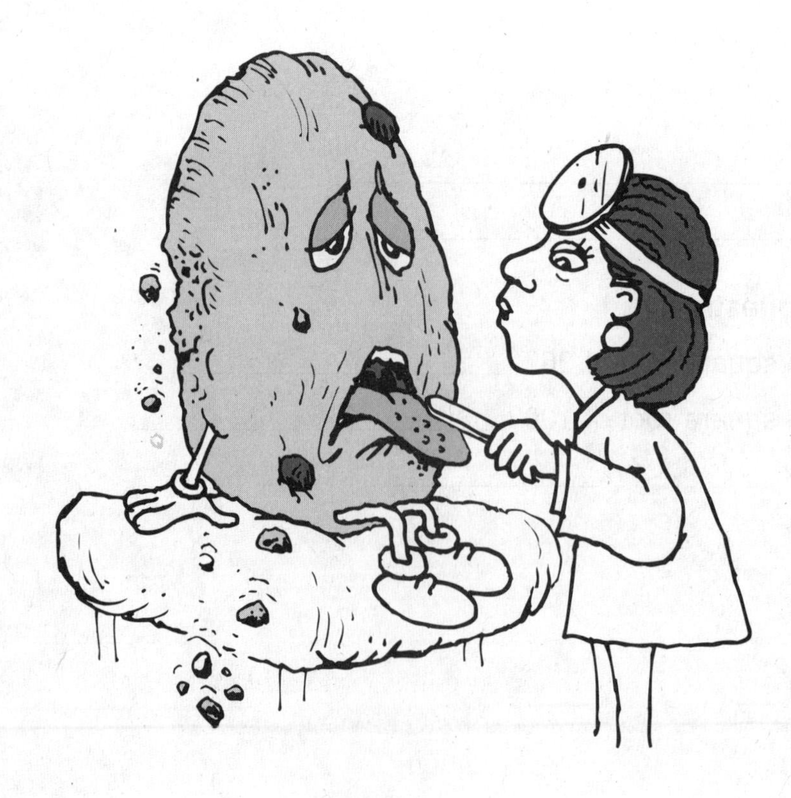

Name ______________________ Date ____________ Class ____________

LESSON **4-6**

Practice A

Estimating Square Roots

Each square root is between two integers. Name the integers. Explain your answer.

1. $\sqrt{10}$

2. $\sqrt{8}$

3. $\sqrt{19}$

4. $\sqrt{33}$

5. $\sqrt{15}$

6. $\sqrt{39}$

Use a calculator to find each value. Round to the nearest tenth.

7. $\sqrt{12}$ **8.** $\sqrt{18}$ **9.** $\sqrt{7}$ **10.** $\sqrt{24}$

11. $\sqrt{38}$ **12.** $\sqrt{45}$ **13.** $\sqrt{8}$ **14.** $\sqrt{22}$

15. $\sqrt{54}$ **16.** $\sqrt{27}$ **17.** $\sqrt{40}$ **18.** $\sqrt{48}$

The distance that a person can see from a certain height to the horizon can be determined by using the formula *View* = $\sqrt{1.6 \cdot \textit{height}}$ where the height is in feet and the result is in statute miles (defines miles between cities).

19. How far to the horizon could you see if standing on top of Pike's Peak in Colorado, elevation 14,110 ft? Round the answer to the nearest tenth.

20. How far to the horizon could you see if flying in a plane at an altitude of 30,000 ft? Round the answer to the nearest tenth.

Name ______________________________ Date ____________ Class ____________

LESSON 4-6

Practice B

Estimating Square Roots

Each square root is between two integers. Name the integers. Explain your answer.

1. $\sqrt{6}$

____

2. $\sqrt{20}$

____

3. $\sqrt{28}$

____

4. $\sqrt{44}$

____

5. $\sqrt{31}$

____

6. $\sqrt{52}$

____

Use a calculator to find each value. Round to the nearest tenth.

7. $\sqrt{14}$ ________ **8.** $\sqrt{42}$ ________ **9.** $\sqrt{21}$ ________ **10.** $\sqrt{47}$ ________

11. $\sqrt{58}$ ________ **12.** $\sqrt{60}$ ________ **13.** $\sqrt{35}$ ________ **14.** $\sqrt{75}$ ________

Police use the formula $r = 2\sqrt{5L}$ to approximate the rate of speed in miles per hours of a vehicle from its skid marks, where L is the length of the skid marks in feet.

15. About how fast is a car going that leaves skid marks of 80 ft?

____

16. About how fast is a car going that leaves skid marks of 245 ft?

____

17. If the formula for finding the length of the skid marks is $L = \frac{r^2}{20}$, what would be the length of the skid marks from a vehicle traveling 80 mi/h?

____

Name ______________________ Date ____________ Class ____________

LESSON **4-6**

Practice C

Estimating Square Roots

Each square root is between two integers. Name the integers. Explain your answer.

1. $\sqrt{75}$

2. $\sqrt{104}$

3. $\sqrt{180}$

4. $\sqrt{230}$

Use a calculator to find each value. Round to the nearest tenth. Use the indicated letter to graph each point on the number line.

5. *A:* $\sqrt{1.21}$ **6.** *B:* $-\sqrt{2.56}$ **7.** *C:* $\sqrt{0.36}$ **8.** *D:* $-\sqrt{0.09}$

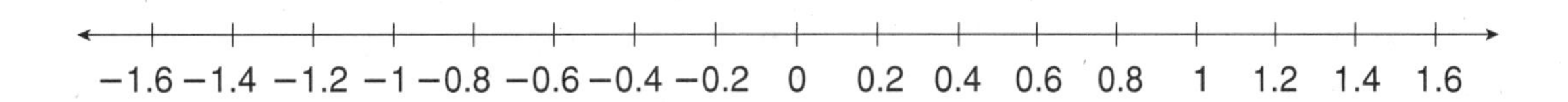

Use guess and check to estimate each square root to the nearest hundredth.

9. $-\sqrt{141}$ **10.** $\sqrt{195}$ **11.** $-\sqrt{0.0245}$ **12.** $\sqrt{47.7}$

Find each product to the nearest hundredth.

13. $\sqrt{64} \cdot \sqrt{24}$ **14.** $-\sqrt{53} \cdot \sqrt{38}$ **15.** $\sqrt{35} \cdot (-\sqrt{16})$ **16.** $\sqrt{25} \cdot \sqrt{28}$

17. $\sqrt{112} \cdot \sqrt{38}$ **18.** $\sqrt{77} \cdot \sqrt{142}$ **19.** $-\sqrt{66} \cdot (-\sqrt{98})$ **20.** $\sqrt{81} \cdot (-\sqrt{217})$

21. If the distance in feet traveled by a falling object is determined by the formula $d = 16t^2$, in which t is the time in seconds, find the distance a falling object traveled in 22 sec.

22. If the distance traveled by the falling object was 3600 ft, how many seconds did it take to fall?

Name ______________________ Date __________ Class __________

LESSON 4-6

Reteach

Estimating Square Roots

To locate a square root between two integers, refer to the table.

Number	1	2	3	4	5	6	7	8	9	10
Square	1	4	9	16	25	36	49	64	81	100
Number	11	12	13	14	15	16	17	18	19	20
Square	121	144	169	196	225	256	289	324	361	400

Locate $\sqrt{260}$ between two integers.

260 is between the perfect squares 256 and 289: $256 < 260 < 289$

So: $\sqrt{256} < \sqrt{260} < \sqrt{289}$

And: $16 < \sqrt{260} < 17$

Use the table to complete the statements.

1. ____ $< 39 <$ ____

____ $< \sqrt{39} <$ ____

____ $< \sqrt{39} <$ ____

2. ____ $< 130 <$ ____

____ $< \sqrt{130} <$ ____

____ $< \sqrt{130} <$ ____

After locating a square root between two integers, you can determine which of the two integers the square root is closer to.

27 is between the perfect squares 25 and 36: $25 < 27 < 36$

So: $\sqrt{25} < \sqrt{27} < \sqrt{36}$

And: $5 < \sqrt{27} < 6$

The difference between 27 and 25 is 2;
the difference between 36 and 27 is 9.
So, $\sqrt{27}$, is closer to 5.

$25 < 27 < 36$

2 9

Complete the statements.

3. $100 < 106 < 121$

____ $< \sqrt{106} <$ ____

____ $< \sqrt{106} <$ ____

$106 - 100 =$ ____

$121 - 106 =$ ____

$\sqrt{106}$ is closer to ____ than ____

4. ____ $< 250 <$ ____

____ $< \sqrt{250} <$ ____

____ $< \sqrt{250} <$ ____

$250 -$ ____ $=$ ____

____ $- 250 =$ ____

$\sqrt{250}$ is closer to ____ than ____

Name ______________________ Date __________ Class __________

LESSON 4-6

Challenge

Dig Deeper!

The **digital root** of a number is found by adding its digits, adding the digits of the result, and so on, until the result is a single digit.

918 ➤ 9 + 1 + 8 = 18 ➤ 1 + 8 = 9   The digital root of 918 is 9.

1. Complete the table to display numbers and their digital roots and to determine if they are divisible by 3 (remainder = 0). Make an observation about the results.

____

Number	Divisible by 3?	Digital Root Calculation			Divisible by 3?
81			=		
92			=		
226			=		
315			=		
659	no	6 + 5 + 9 = 20 ➤ 2 + 0	=	2	no
704			=		
1064			=		

2. Complete the table to display the products of numbers and the products of their digital roots. Make an observation about the results.

____

____

Product	Digital Root of Factor	Digital Root of Factor	Product of Digital Roots of Factors	Digital Root of Product
24 × 32 = 768	2 + 4 = 6	3 + 2 = 5	6 × 5 = 30 ➤ 3 + 0 = 3	7 + 6 + 8 = 21 ➤ 2 + 1 = 3
11 × 17 =				
121 × 42 =				
243 × 35 =				
81 × 72 =				
360 × 54 =				

Name ______________________ Date __________ Class __________

LESSON 4-6

Problem Solving

Estimating Square Roots

The distance to the horizon can be found using the formula $d = 112.88\sqrt{h}$ where d is the distance in kilometers and h is the number of kilometers from the ground. Round your answer to the nearest kilometer.

1. How far is it to the horizon when you are standing on the top of Mt. Everest, a height of 8.85 km?

____

2. Find the distance to the horizon from the top of Mt. McKinley, Alaska, a height of 6.194 km.

____

3. How far is it to the horizon if you are standing on the ground and your eyes are 2 m above the ground?

____

4. Mauna Kea is an extinct volcano on Hawaii that is about 4 km tall. You should be able to see the top of Mauna Kea when you are how far away?

____

You can find the approximate speed of a vehicle that leaves skid marks before it stops. The formulas $S = 5.5\sqrt{0.7L}$ and $S = 5.5\sqrt{0.8L}$, where S is the speed in miles per hour and L is the length of the skid marks in feet, will give the minimum and maximum speeds that the vehicle was traveling before the brakes were applied. Round to the nearest mile per hour.

5. A vehicle leaves a skid mark of 40 feet before stopping. What was the approximate speed of the vehicle before it stopped?

A 25–35 mph **C** 29–31 mph
B 28–32 mph **D** 68–70 mph

6. A vehicle leaves a skid mark of 100 feet before stopping. What was the approximate speed of the vehicle before it stopped?

F 46–49 mph **H** 62–64 mph
G 50–55 mph **J** 70–73 mph

7. A vehicle leaves a skid mark of 150 feet before stopping. What was the approximate speed of the vehicle before it stopped?

A 50–55 mph **C** 55–70 mph
B 53–58 mph **D** 56–60 mph

8. A vehicle leaves a skid mark of 200 feet before stopping. What was the approximate speed of the vehicle before it stopped?

F 60–63 mph **G** 65–70 mph
H 72–78 mph **J** 80–90 mph

Name ______________________________ Date __________ Class __________

LESSON 4-6

Reading Strategies

Follow a Procedure

The numbers 16 and 25 are called **perfect squares.** Each has an integer as its square root. To find the square root of a perfect square, ask yourself what number multiplied by itself equals the perfect square.

Some Perfect Squares				
1	4	9	16	25
36	49	64	81	
100	121	144	169	

1. What number times itself equals 16? ____________________

2. What is the square root of 16? ____________________

3. What number times itself equals 25? ____________________

4. What is the square root of 25? ____________________

Use these steps to estimate the square root of a number that is not a perfect square.

What is $\sqrt{45}$?

Step 1
Identify a perfect square that is a little more than $\sqrt{45}$. → $\sqrt{49}$
The square root of 49 = 7.

Step 2
Identify a perfect square that is a little less than $\sqrt{45}$. → $\sqrt{36}$
The square root of 36 = 6.

Step 3
The estimate of $\sqrt{45}$ is between 6 and 7.

Use the steps above to help you estimate the square root of 90.

5. Which perfect square is a little more than 90? __________

6. What is the square root of 100? __________

7. Which perfect square is a little less than 90? __________

8. What is the square root of 81? __________

9. What is your estimate of the square root of 90?

____

Name ______________________ Date __________ Class __________

LESSON 4-6

Puzzles, Twisters & Teasers

The Root of the Problem!

Find the square roots. Use the answers to solve the riddle.

S $\sqrt{36}$ = __________

R $\sqrt{144}$ = __________

P $\sqrt{100}$ = __________

T $\sqrt{64}$ = __________

G $\sqrt{25}$ = __________

I $\sqrt{9}$ = __________

W $\sqrt{4}$ = __________

H $\sqrt{169}$ = __________

E $\sqrt{81}$ = __________

U $\sqrt{49}$ = __________

L $\sqrt{121}$ = __________

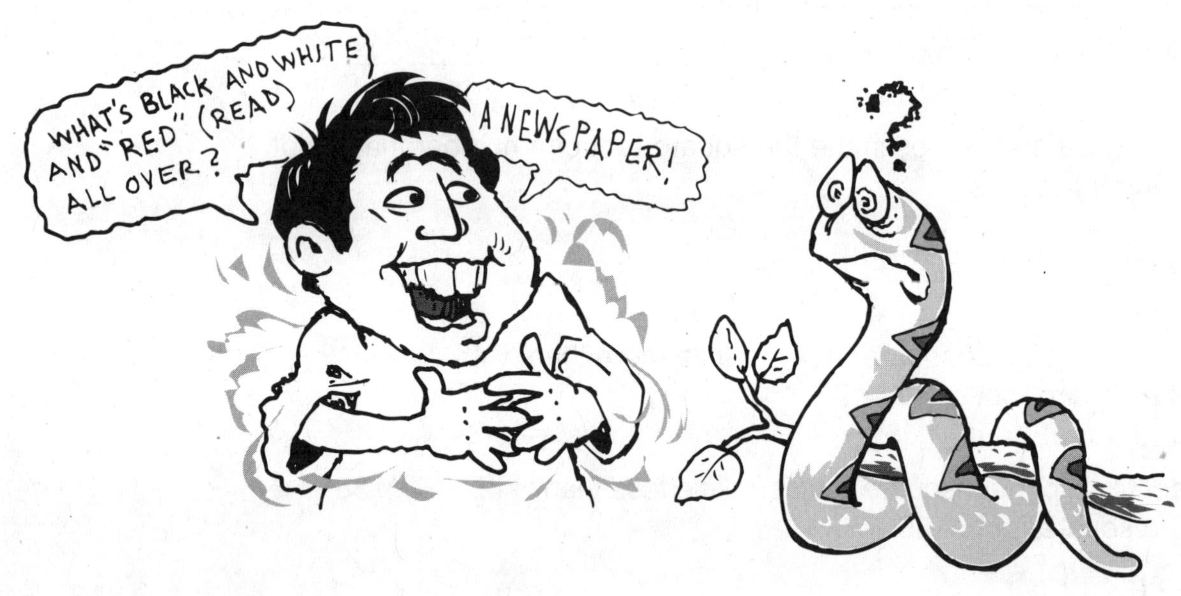

Why can't you play jokes on snakes?

Because you can't ___ ___ ___ ___
10 7 11 11

___ ___ ___ ___ ___   ___ ___ ___ ___.
8 13 9 3 12   11 9 5 6

Name ______________________ Date __________ Class __________

LESSON 4-7

Practice A

The Real Numbers

Write all names that apply to each number.

1. -3.2 ______ ______

2. $\frac{2}{5}$ ______ ______

3. $-\sqrt{12}$ ______ ______

4. $\frac{\sqrt{4}}{2}$ ______ ______

5. -20 ______ ______

6. $\sqrt{16}$ ______ ______

State if the number is rational, irrational, or not a real number.

7. 0 ______

8. $\sqrt{7}$ ______

9. $\sqrt{-4}$ ______

10. $\frac{\sqrt{9}}{0}$ ______

11. $\frac{3}{4}$ ______

12. $\sqrt{25}$ ______

13. $-\sqrt{49}$ ______

14. $\sqrt{11}$ ______

Find a real number between each pair of numbers.

15. $3\frac{1}{3}$ and $3\frac{2}{3}$ ______

16. 2.16 and $\frac{11}{5}$ ______

17. $\frac{1}{8}$ and $\frac{1}{5}$ ______

18. Give an example of an irrational number that is greater than 0.

______

19. Give an example of a number that is not real.

______

20. Give an example of a rational number between $\frac{1}{2}$ and $\sqrt{1}$.

______

Name ______________________ Date ____________ Class ____________

LESSON 4-7

Practice B

The Real Numbers

Write all names that apply to each number.

1. $-\frac{7}{8}$

2. $\sqrt{0.15}$

3. $\sqrt{\frac{18}{2}}$

4. $\sqrt{45}$

5. -25

6. -6.75

State if the number is rational, irrational, or not a real number.

7. $\sqrt{14}$

8. $\sqrt{-16}$

9. $\frac{6.2}{0}$

10. $\sqrt{49}$

11. $\frac{7}{20}$

12. $-\sqrt{81}$

13. $\sqrt{\frac{7}{9}}$

14. -1.3

Find a real number between each pair of numbers.

15. $7\frac{3}{5}$ and $7\frac{4}{5}$

16. 6.45 and $\frac{13}{2}$

17. $\frac{7}{8}$ and $\frac{9}{10}$

18. Give an example of a rational number between $-\sqrt{4}$ and $\sqrt{4}$

19. Give an example of an irrational number less than 0.

20. Give an example of a number that is not real.

Name ______________________ Date ____________ Class ____________

LESSON **4-7**

Practice C

The Real Numbers

Write all names that apply to each number.

1. $0.31\overline{8}$

____

____

2. $\frac{\sqrt{17}}{21}$

____

____

3. $\sqrt{-100}$

____

____

4. $\frac{\sqrt{144}}{-6}$

____

____

5. $\sqrt{6.15}$

____

____

6. $-\frac{21}{22}$

____

____

7. $-\sqrt{196}$

____

____

8. $\sqrt{64} + 15$

____

____

9. $\sqrt{112}$

____

____

Use the indicated letter to approximate each irrational number on the number line.

12. *C:* $\sqrt{7}$

13. *D:* $\sqrt{18}$

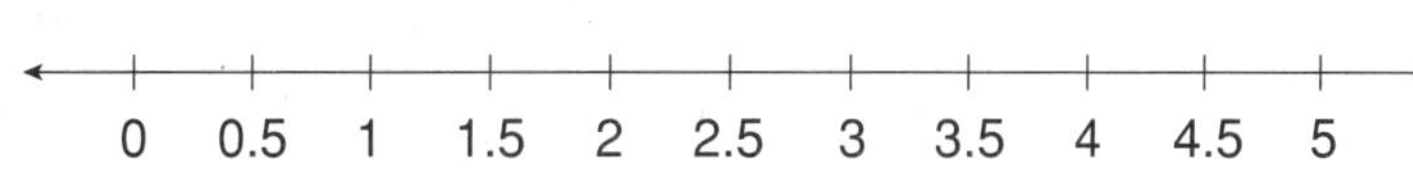

14. Find a real number between $\sqrt{1}$ and $\sqrt{2}$.

____

15. Find a real number between $\frac{11}{12}$ and 1.

____

16. Find a real number between -0.25 and -0.3.

____

17. Find a real number between $\frac{9}{11}$ and $\frac{10}{11}$.

____

18. Find a real number between $\sqrt{28}$ and $\sqrt{30}$.

____

Name ______________________ Date __________ Class __________

LESSON 4-7

Reteach

The Real Numbers

The set of **rational numbers** contains all integers, all fractions, and decimals that end or repeat.

Irrational numbers can only be written as decimals that do not end or repeat.

Together, the rational numbers and the irrational numbers form the set of **real numbers.**

Real Numbers → Rational Numbers, Irrational Numbers

Square roots of numbers that are perfect squares are rational.

$\sqrt{25} = 5$

Square roots of numbers that are not perfect squares are irrational.

$\sqrt{3} = 1.732050807...$

Tell if each number is rational or irrational.

1. $\sqrt{7}$ __________ **2.** $\sqrt{81}$ __________ **3.** $\sqrt{169}$ __________ **4.** $\sqrt{101}$ __________

The square of a nonzero number is positive. $3^2 = 9$ and $(-3)^2 = 9$
So, the square root of a negative number is not a real number.

$\sqrt{-9}$ is not a real number.

Tell if each number is real or not real.

5. -8 __________ **6.** $-\sqrt{8}$ __________ **7.** $\sqrt{-8}$ __________ **8.** $\sqrt{-25}$ __________

Between any two real numbers, there is always another real number. One way to find a number between is to find the number halfway between.

To find a real number between $7\frac{1}{5}$ and $7\frac{2}{5}$,
divide their sum by 2: $7\frac{1}{5} + 7\frac{2}{5} = \left(14\frac{3}{5}\right) \div 2 = 7\frac{3}{10}$

Find a real number between each pair.

9. $8\frac{3}{7}$ and $8\frac{4}{7}$ __________ **10.** -1.6 and -1.7 __________ **11.** $-3\frac{7}{9}$ and $-3\frac{2}{9}$ __________ **12.** $6\frac{1}{2}$ and $6\frac{3}{4}$ __________

Name ______________________ Date __________ Class __________

LESSON 4-7

Challenge

Searching for Perfection

Numbers that are equal to the sum of all their factors (not including the number itself) are called **perfect numbers.**

$6 = 1 + 2 + 3$ 6 is the smallest perfect number.

1. Which of the numbers 24 or 28 is a perfect number? Explain.

____

____

The ancient Greek mathematician Euclid devised a method for computing perfect numbers.

- Begin with the number 1 and keep adding powers of 2 until you get a sum that is a *prime number* (only factors are itself and 1).
- Multiply this sum by the last power of 2.

2. Complete the table to write the first three perfect numbers.

	Sum	Prime?	Euclid's Method	Perfect Number
1 + 2	= 3	yes	2 × 3	6
1 + 2 + 4	=			
1 + 2 + 4 + 8	=			
	=			

So the first three perfect numbers are ______________.

The next perfect number is tedious to calculate in this manner. If, however, the calculations are written with exponents, a new pattern emerges.

3. Complete the table to write the sums using exponents.

Series	Sum
$1 + 2^1$	$= 2^2 - 1$
$1 + 2^1 + 2^2$	$= 2^3 - 1$
$1 + 2^1 + 2^2 + 2^3$	=
$1 + 2^1 + 2^2 + 2^3 + 2^4$	=

Incorporating this information, Euclid proved that whenever a prime number of the form $2^n - 1$ is found, a perfect number can be written.

If $2^n - 1$ is prime, then $2^{n-1}(2^n - 1)$ is a perfect number.

4. Find the fourth perfect number. ______________________

5. Find the fifth perfect number. ______________________

Name ______________________ Date ____________ Class ____________

LESSON 4-7

Problem Solving

The Real Numbers

Write the correct answer.

1. Twin primes are prime numbers that differ by 2. Find an irrational number between twin primes 5 and 7.

____

2. Rounded to the nearest ten-thousandth, $\pi = 3.1416$. Find a rational number between 3 and π.

____

3. One famous irrational number is *e*. Rounded to the nearest ten-thousandth $e \approx 2.7183$. Find a rational number that is between 2 and *e*.

____

4. Perfect numbers are those that the divisors of the number sum to the number itself. The number 6 is a perfect number because $1 + 2 + 3 = 6$. The number 28 is also a perfect number. Find an irrational number between 6 and 28.

____

Choose the letter for the best answer.

5. Which is a rational number?

A the length of a side of a square with area 2 cm^2

B the length of a side of a square with area 4 cm^2

C a non-terminating decimal

D the square root of a prime number

6. Which is an irrational number?

F a number that can be expressed as a fraction

G the length of a side of a square with area 4 cm^2

H the length of a side of a square with area 2 cm^2

J the square root of a negative number

7. Which is an integer?

A the number half-way between 6 and 7

B the average rainfall for the week if it rained 0.5 in., 2.3 in., 0 in., 0 in., 0 in., 0.2 in., 0.75 in. during the week

C the money in an account if the balance was $213.00 and $21.87 was deposited

D the net yardage after plays that resulted in a 15 yard loss, 10 yard gain, 6 yard gain and 5 yard loss

8. Which is a whole number?

F the number half-way between 6 and 7

G the total amount of sugar in a recipe that calls for $\frac{1}{4}$ cup of brown sugar and $\frac{3}{4}$ cup of granulated sugar

H the money in an account if the balance was $213.00 and $21.87 was deposited

J the net yardage after plays that resulted in a 15 yard loss, 10 yard gain, 6 yard gain and 5 yard loss

Name ______________________ Date ____________ Class ____________

LESSON **4-7**

Reading Strategies

Use a Venn Diagram

You know that **rational numbers** can be written in fraction form as an $\frac{\text{integer}}{\text{integer}}$. Rational numbers include:

- **Decimals**
- **Fractions**
- **Integers**
- **Whole Numbers**

This diagram of rational numbers expressed in different forms helps you see how they are related.

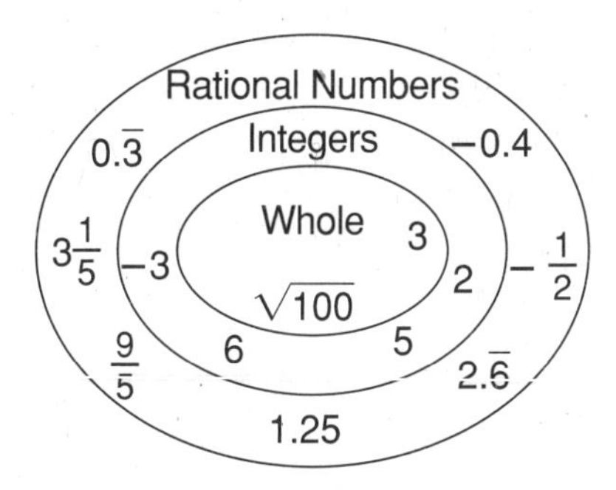

From this picture you can say:

1. -0.4 is a rational number, but it is not an integer or ____________
2. $\sqrt{100} = 10$. It is a rational number, it is ____________, and it is a whole number.
3. -3 is a rational number and an integer, but it is not ____________.
4. $2.\overline{6}$ is a rational number, but it is not ____________ or a whole number.

Numbers that are not rational are called **irrational numbers**. For example, $\sqrt{3}$ is an irrational number. It is a decimal that does not terminate or repeat. $\sqrt{3} = 1.7320508\ldots$

Write all names that apply to each number: rational, irrational, integer, or whole number.

5. $2.236068\ldots$ ____________________.
6. -7 ____________________.
7. 328 ____________________.
8. $2\frac{2}{3}$ ____________________.

Name ______________________ Date __________ Class __________

LESSON **4-7**

Puzzles, Twisters & Teasers

Get Real!

Circle the correct word to complete the sentences. Then use the question numbers and answer letters as your key to unlocking the answer to the riddle.

1. __________ numbers can be written as fractions or as decimals that either terminate or repeat.

S	**T**
Rational	Irrational

2. __________ numbers can only be written as decimals that do not terminate or repeat.

B	**D**
Rational	Irrational

3. A repeating decimal will __________ appear to repeat on a calculator.

K	**L**
sometimes	always

4. The set of __________ numbers consists of the set of rational numbers and the set of irrational numbers.

E	**O**
whole	real

5. The square root of a negative number __________ a real number.

S	**C**
is	is not

6. The density property of real numbers states that between any two real numbers is another __________ number.

M	**N**
unreal	real

7. There __________ an integer between −2 and −3.

P	**L**
is	is not

8. 0.25 would be an example of a __________ decimal.

S	**T**
repeating	terminating

9. Ten is a whole number that __________ a perfect square.

O	**E**
is	is not

10. A number and its reciprocal have a product of __________.

P	**B**
one	zero

Why are pianos hard to open?

Because the keys ___ ___ ___ ___
2 4 6 8

___ ___ ___ ___   ___ ___ ___ ___ ___.
4 10 9 6   7 4 5 3 1

Name ______________________ Date __________ Class __________

LESSON **4-8**

Practice A

The Pythagorean Theorem

Find the length of the hypotenuse in each triangle using the Pythagorean Theorem, $a^2 + b^2 = c^2$.

1.

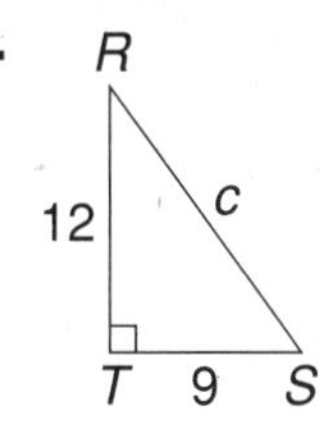

____

2.

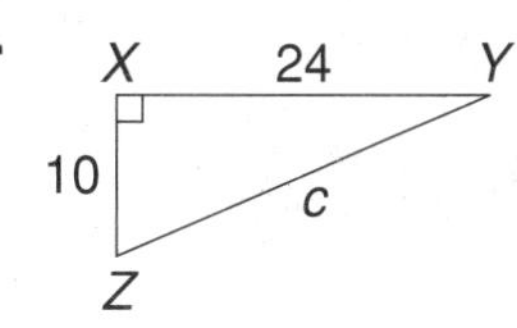

____

3.

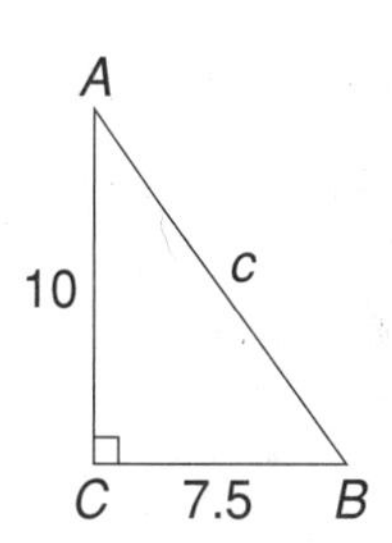

____

Solve for the unknown side in each right triangle. Round the answers to the nearest tenth.

4.

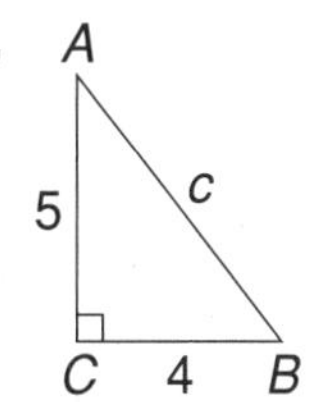

____

5.

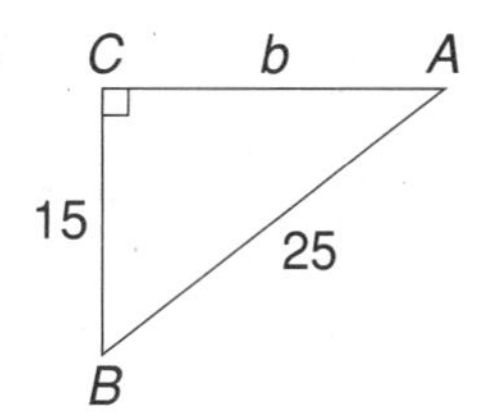

____

6.

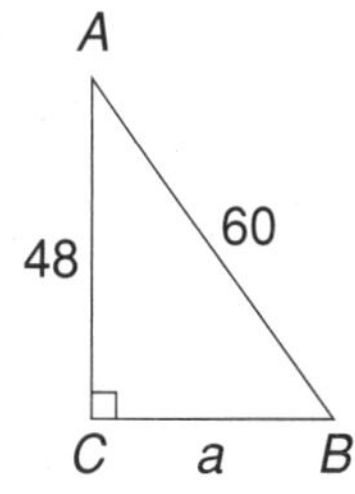

____

7.

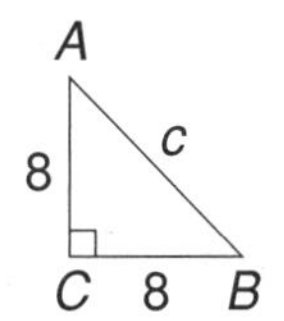

____

8.

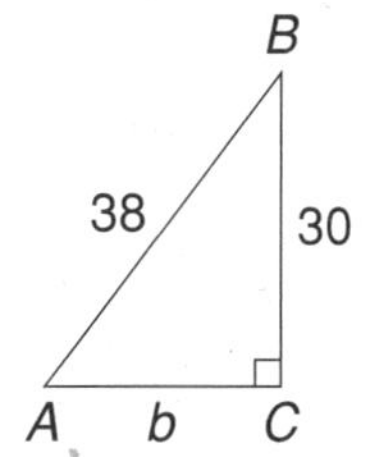

____

9.

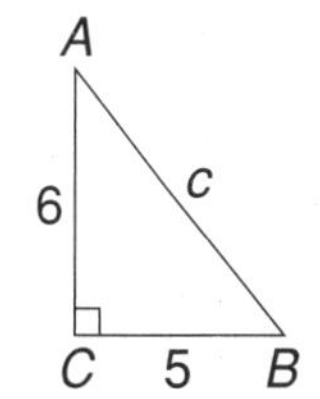

____

10. Jan and her brother Mel go to different schools. Jan goes 6 kilometers east from home. Mel goes 8 kilometers north. How many kilometers apart are their schools?

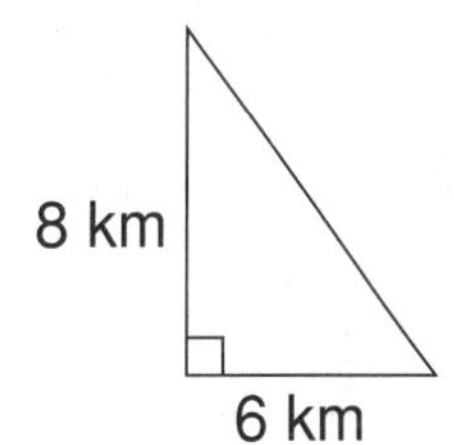

____

Name ______________________ Date __________ Class __________

LESSON 4-8

Practice B

The Pythagorean Theorem

Find the length of the hypotenuse to the nearest tenth.

1.

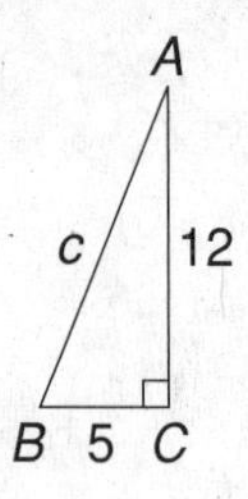

____

2.

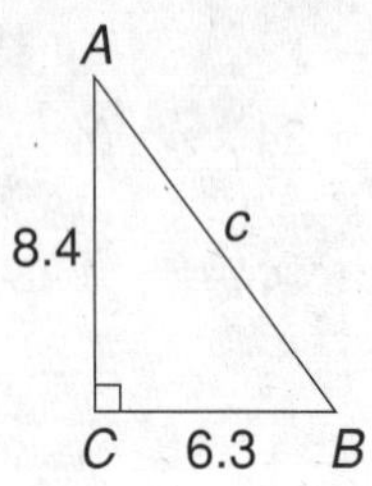

____

3.

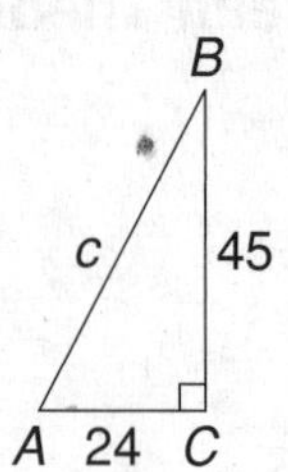

____

Solve for the unknown side in each right triangle to the nearest tenth.

4.

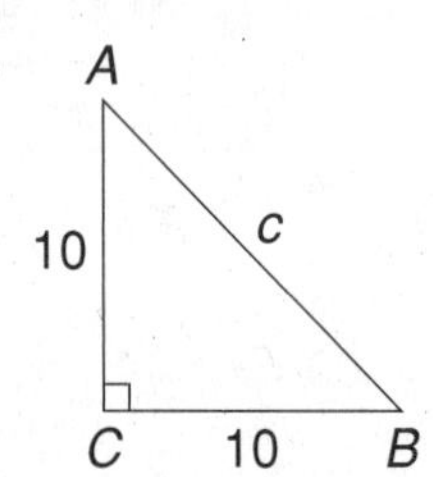

____

5.

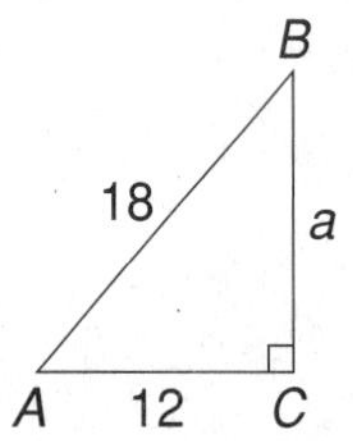

____

6.

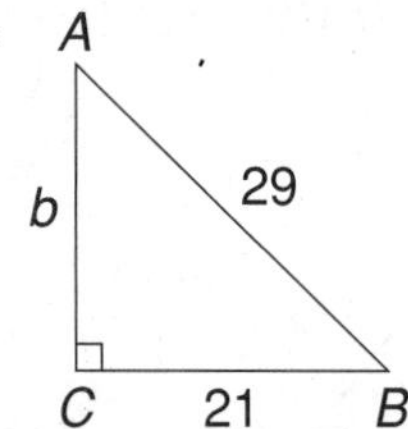

____

7.

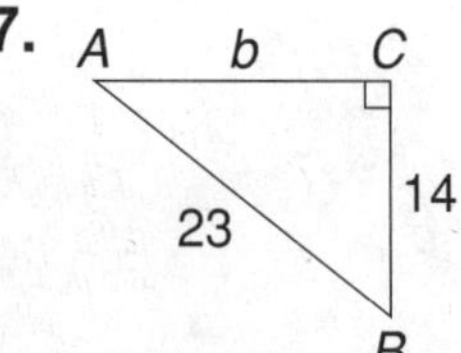

____

8.

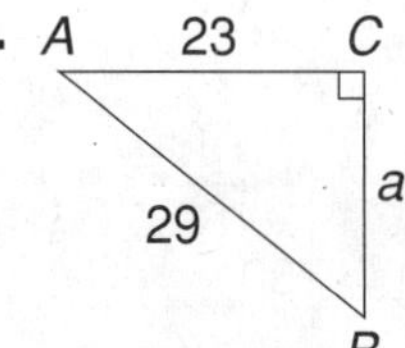

____

9.

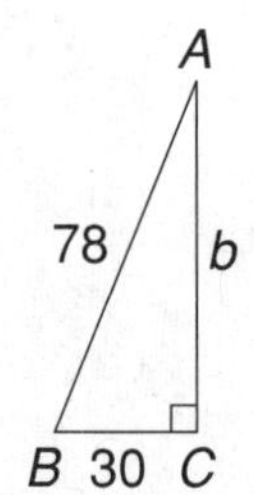

____

10. A glider flies 8 miles south from the airport and then 15 miles east. Then it flies in a straight line back to the airport. What was the distance of the glider's last leg back to the airport?

____

Name ______________________________ Date ____________ Class ____________

LESSON **4-8**

Practice C

The Pythagorean Theorem

Solve for the unknown side in each right triangle to the nearest tenth..

1.

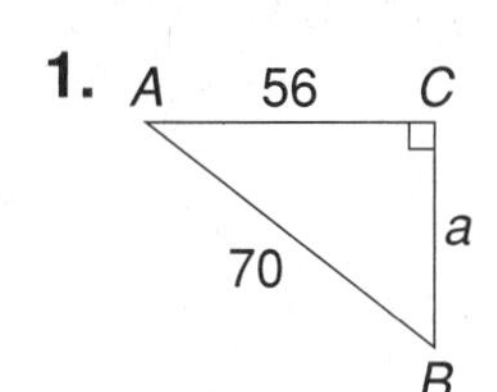

2.

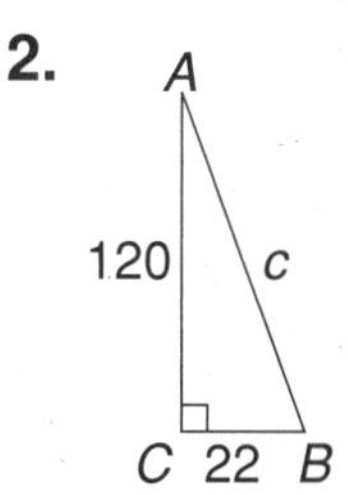

3.

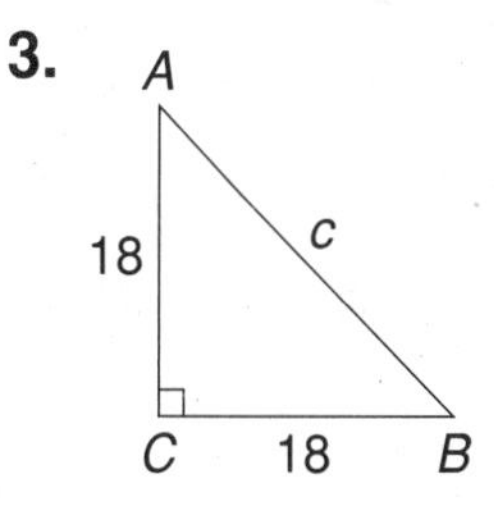

____________ ____________ ____________

4. $a = 8$, $b = 15$, $c = ?$

____

5. $a = 0.5$, $b = ?$, $c = 1.3$

____

6. $a = ?$, $b = 18$, $c = 28$

____

7. $a = 21$, $b = ?$, $c = 46$

____

8. $a = ?$, $b = 38$, $c = 45$

____

9. $a = 30$, $b = ?$, $c = 50$

____

10. $a = 30$, $b = 72$, $c = ?$

____

11. $a = 40$, $b = ?$, $c = 65$

____

12. $a = 65$, $b = ?$, $c = 97$

____

Determine whether each set is a Pythagorean triple.

13. 2.1, 2.8, 3.5

____

14. 12, 15, 20

____

15. 30, 70, 78

____

16. 18, 24, 30

____

17. Use the Pythagorean Theorem to find the missing side of the triangle if the hypotenuse is 68 and the other side is 32.

____

18. Use the Pythagorean Theorem to find the base of this triangle.

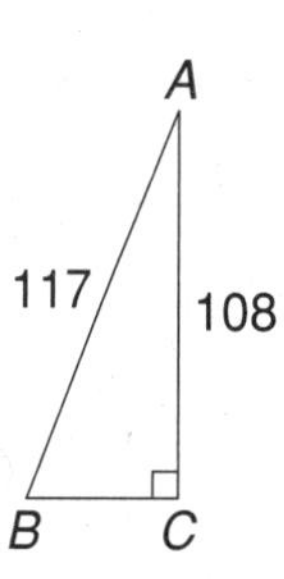

____

19. A 20-ft ladder is leaning against a house. The bottom of the ladder is 3 ft from the house. To the nearest tenth of a foot, about how high does the top of the ladder reach?

____

Name ______________________ Date __________ Class __________

LESSON 4-8

Reteach

The Pythagorean Theorem

In a right triangle,
the sum of the areas of the squares on the legs
is equal to
the area of the square on the hypotenuse.

$3^2 + 4^2 = 5^2$

$9 + 16 = 25$

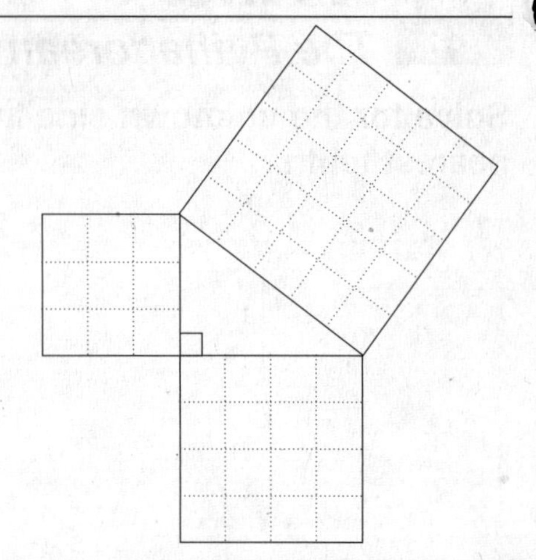

Given the squares that are on the legs of a right triangle, draw the square for the hypotenuse.

1. leg leg hypotenuse

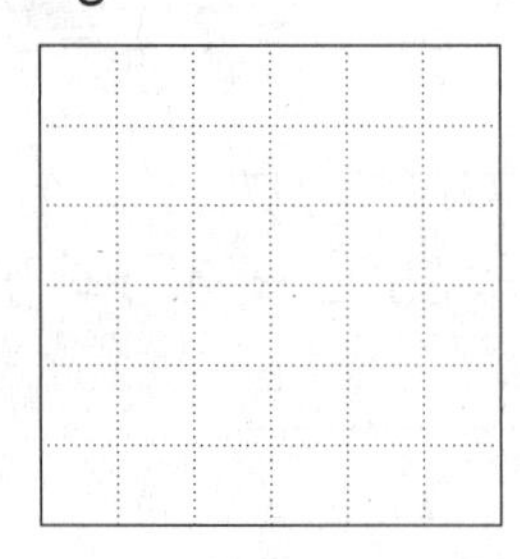

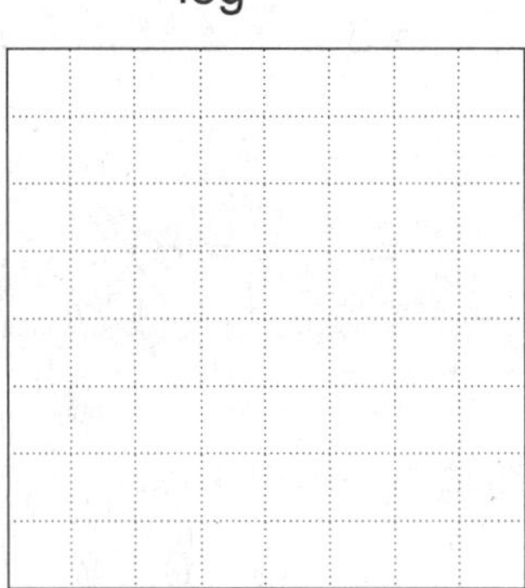

Without drawing the squares, you can find the length of a side.

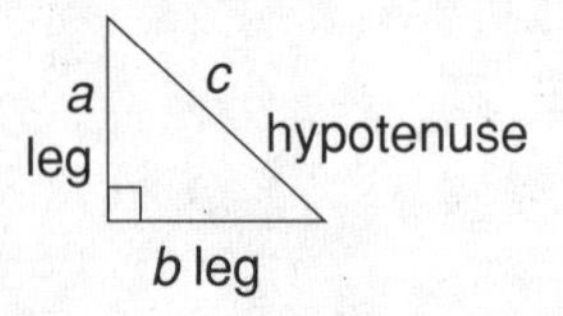

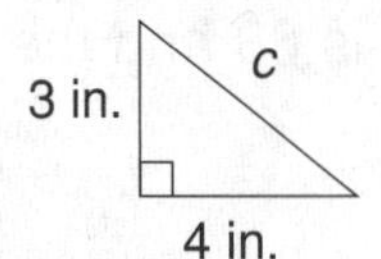

$a^2 + b^2 = c^2$
$3^2 + 4^2 = c^2$
$9 + 16 = c^2$
$25 = c^2$
$c = 5$ in.

Complete to find the length of each hypotenuse.

2. 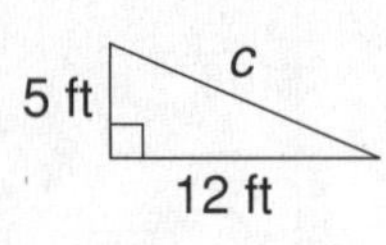

$a^2 + b^2 = c^2$

_____ + _____ $= c^2$

_____ + _____ $= c^2$

_____ $= c^2$

$c =$ _____ ft

3. 
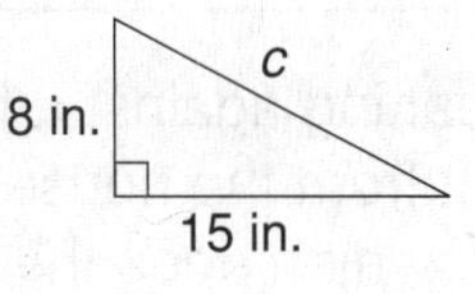

$a^2 + b^2 = c^2$

_____ + _____ $= c^2$

_____ + _____ $= c^2$

_____ $= c^2$

$c =$ _____ in.

Name ______________________ Date __________ Class __________

LESSON 4-8

Reteach

The Pythagorean Theorem (continued)

You can use the Pythagorean Theorem to find the length of a leg if you know the length of the other leg and the hypotenuse.

15 in.
a
12 in.

$$a^2 + b^2 = c^2$$

$a^2 + \_\_\_\_ = \_\_\_\_$

$a^2 + \_\_\_\_ = \_\_\_\_$

$- \_\_\_\_ \quad - \_\_\_\_$

$a^2 = \_\_\_\_$

$a = \_\_\_\_$ in.

Complete to find the length of each leg.

4. $a^2 + b^2 = c^2$

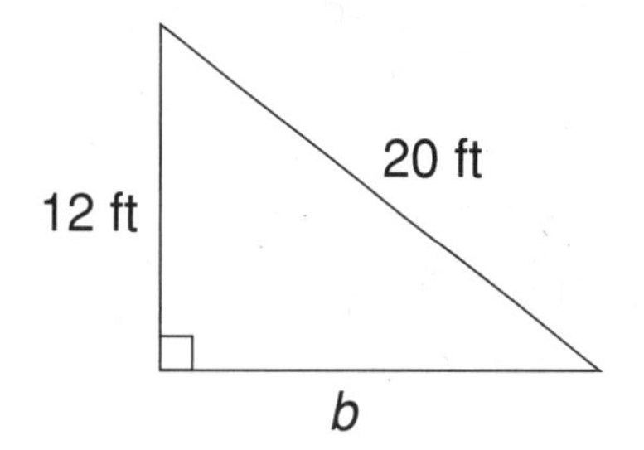

$\_\_\_\_ + b^2 = \_\_\_\_$

$\_\_\_\_ + b^2 = \_\_\_\_$

$- \_\_\_\_ \quad - \_\_\_\_$

$b^2 = \_\_\_\_$

$b = \_\_\_\_$ in.

5. $a^2 + \_\_\_\_ = \_\_\_\_$

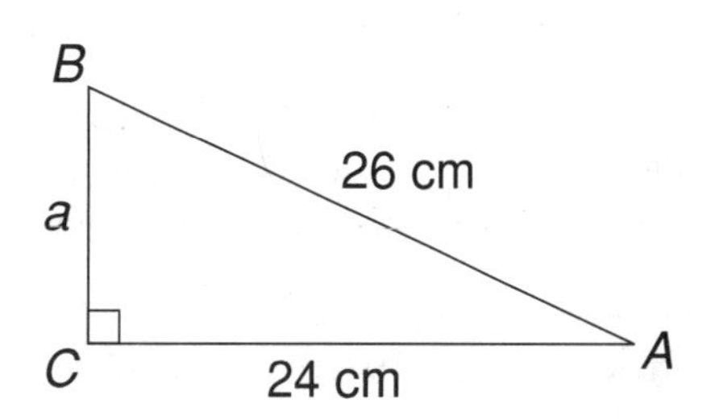

$a^2 + \_\_\_\_ = \_\_\_\_$

$- \_\_\_\_ \quad - \_\_\_\_$

$a^2 = \_\_\_\_$

$a = \_\_\_\_$ cm

Name ______________________ Date ____________ Class ____________

LESSON 4-8

Challenge

Triple Play

Three numbers connected by the Pythagorean relation are called **Pythagorean triples.**

Since $3^2 + 4^2 = 5^2$, the numbers 3-4-5 are a Pythagorean Triple.

Consider the Pythagorean triples shown in the table.

	Column *A*	Column *B*	Column *C*
row 1	3	4	5
row 2	5	12	13
row 3	7	24	25
row 4	9	40	41
row 5	11	60	61

1. Make an observation about the numbers in Column *A*.

   ____

2. How are the numbers in Column *C* related to those in Column *B*?

   ____

3. Complete this table by carrying out the indicated calculation. Two calculations are done.

   Compare the results to the Pythagorean triples in Columns *A*, *B*, and *C* of the original table.

   ____

	Column *A*	row × *A* + row
row 1	3	$1 \times 3 + 1 = 4$
row 2	5	$2 \times 5 + 2 = 12$
row 3	7	
row 4	9	
row 5	11	

4. In the original table, how do the squares of the numbers in Column *A* relate to the numbers in Columns *B* and *C*?

   ____

5. Using the relationships you have observed, calculate rows 6 and 10 of the table of Pythagorean triples. Verify your results by applying the Pythagorean Theorem.

	Column *A*	Column *B*	Column *C*	Verify $A^2 + B^2 = C^2$
row 6				
row 10				

Name ______________________ Date ____________ Class ____________

LESSON **4-8**

Problem Solving

The Pythagorean Theorem

Write the correct answer. Round to the nearest tenth.

1. A utility pole 10 m high is supported by two guy wires. Each guy wire is anchored 3 m from the base of the pole. How many meters of wire are needed for the guy wires?

____

2. A 12 foot-ladder is resting against a wall. The base of the ladder is 2.5 feet from the base of the wall. How high up the wall will the ladder reach?

____

3. The base-path of a baseball diamond form a square. If it is 90 ft from home to first, how far does the catcher have to throw to catch someone stealing second base?

____

4. A football field is 100 yards with 10 yards at each end for the end zones. The field is 45 yards wide. Find the length of the diagonal of the entire field, including the end zones.

____

Choose the letter for the best answer.

5. The frame of a kite is made from two strips of wood, one 27 inches long, and one 18 inches long. What is the perimeter of the kite? Round to the nearest tenth.

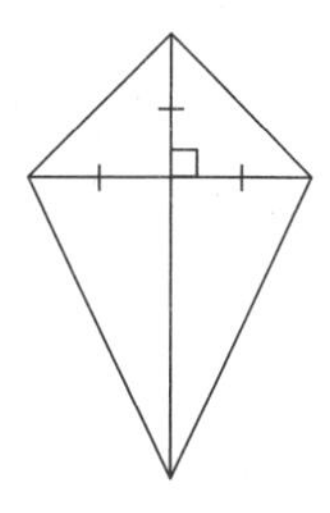

A 18.8 in. **C** 65.7 in.

B 32.8 in. **D** 131.2 in.

6. The glass for a picture window is 8 feet wide. The door it must pass through is 3 feet wide. How tall must the door be for the glass to pass through the door? Round to the nearest tenth.

F 3.3 ft **H** 7.4 ft

G 6.7ft **J** 8.5 ft

7. A television screen measures approximately 15.5 in. high and 19.5 in. wide. A television is advertised by giving the approximate length of the diagonal of its screen. How should this television be advertised?

A 25 in. **C** 12 in.

B 21 in. **D** 6 in.

8. To meet federal guidelines, a wheelchair ramp that is constructed to rise 1 foot off the ground must extend 12 feet along the ground. How long will the ramp be? Round to the nearest tenth.

F 11.9 ft **H** 13.2 ft

G 12.0 ft **J** 15.0 ft

Name ______________________________ Date ____________ Class ____________

LESSON 4-8

Reading Strategies

Vocabulary

Right triangles are a special type of triangle. They have one **right angle.** The right angle measures 90°. The angle may be marked with a small square.

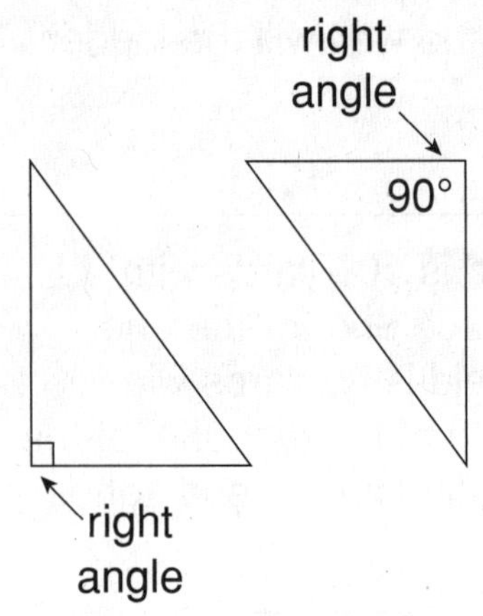

The two shorter sides that form the right angle of a triangle are called **legs.** The third and longest side of a right triangle is called the **hypotenuse.** The hypotenuse is always located opposite the right angle.

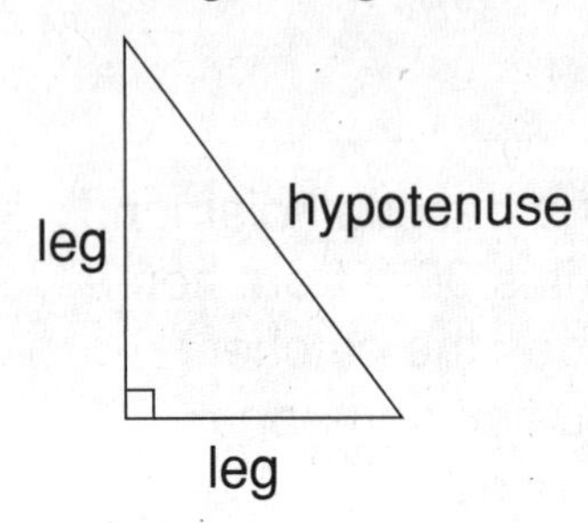

Answer each question.

1. How many right angles does a right triangle have?

____

2. What name is given to the two sides of the triangle that form the right angle?

____

3. What symbol is used to identify a right angle?

____

4. How many degrees are in a right angle?

____

5. What name is given to the longest side of a right triangle?

____

6. What is the name of the side opposite the right angle?

____

7. Is it possible for a right triangle to have two right angles? Why or why not?

____

____

Name ______________________ Date __________ Class __________

LESSON **4-8**

Puzzles, Twisters & Teasers

The Root of the Problem!

Find the length of the hypotenuse in each triangle. Each answer has a corresponding letter. Use the letters to solve the riddle.

1. T = ______

T 1 1

2. Y = ______

Y 8 7

3. D = ______

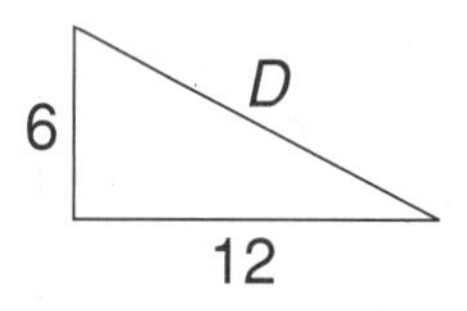

4. N = ______

6 *N* 8

5. A = ______

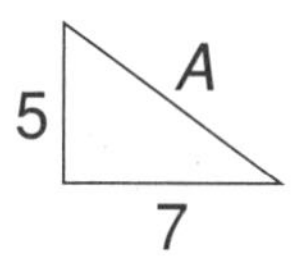

6. I = ______

5 *I* 2

If a dog is tied to a rope 15 feet long, how can it reach a bone 30 feet away? The rope isn't

T ______ E ______ TO ______ N ______ ______ H I ______ G

5.4 13.4 8.6 10.6 1.4 10

LESSON 4-1
Practice A
Exponents

Name the base and exponent for each power.

1. 15^3 base 15 exponent 3
2. 8^{10} base 8 exponent 10
3. $(-3)^2$ base −3 exponent 2

Write using exponents.

4. 2 • 2 • 2 — 2^3
5. 3 • 3 • 3 • 3 • 3 • 3 — 3^6
6. (−1) • (−1) • (−1) • (−1) — $(-1)^4$
7. (−5) • (−5) — $(-5)^2$
8. $a \cdot a \cdot a \cdot a \cdot a$ — a^5
9. (−2) • (−2) • (−2) • (−2) — $(-2)^4$

Evaluate.

10. 3^2 — 9
11. 2^4 — 16
12. 4^3 — 64
13. $(-1)^4$ — 1
14. $(-2)^3$ — −8
15. $(-1)^5$ — −1
16. 5^2 — 25
17. $(-3)^4$ — 81

Evaluate each expression for x = −2 and y = 3.

18. $y^3 + 1$ — 28
19. $x^2 - 1$ — 3
20. $x^4 + y$ — 19
21. $2x - y^2$ — −13

22. Find the area of square with a side of 6 m. (Hint: Area = s^2.)
$36\ m^2$

23. Write an expression for 4 as a factor 5 times.
4^5

/23

LESSON 4-1
Practice B
Exponents

Write in exponential form.

1. 6 • 6 • 6 • 6 • 6 • 6 — 6^6
2. 7 • 7 • 7 • 7 — 7^4
3. (−8) • (−8) • (−8) • (−8) — $(-8)^4$
4. $5 \cdot 5 \cdot 5 \cdot b \cdot b \cdot b \cdot b$ — $5^3 b^4$

Evaluate.

5. 10^2 — 100
6. $(-6)^2$ — 36
7. 8^2 — 64
8. $(-7)^2$ — 49
9. $(-5)^3$ — −125
10. 12^2 — 144
11. $(-9)^2$ — 81
12. $(-4)^3$ — −64
13. 2^5 — 32
14. 5^4 — 625
15. $(-3)^4$ — 81
16. 6^3 — 216

Evaluate each expression for the given values of the variables.

17. $n^3 - 5$ for $n = 4$ — 59
18. $4x^2 + y^3$ for x = 5 and y = −2 — 92
19. $m^p + q^2$ for $m = 5$, $p = 2$, and $q = 4$ — 41
20. $a^4 + 2(b - c^2)$ for $a = 2$, $b = 4$, and $c = -1$ — 22

21. Write an expression for five times a number used as a factor three times.
$5x^3$ or $(5x)^3$

22. Find the volume of a regular cube if the length of a side is 10 cm. (Hint: $V = l^3$.)
$1000\ cm^3$

/22

LESSON 4-1
Practice C
Exponents

Evaluate.

1. $(-10)^2$ — 100
2. 7^3 — 343
3. $(-14)^2$ — 196
4. $-(16)^2$ — −256
5. $(-2)^5$ — −32
6. $(-5)^4$ — 625
7. 3^5 — 243
8. $-(4)^4$ — −256

Evaluate each expression for the given values of the variables.

9. $n^3 - 5p$ for $n = 4$ and $p = 3$ — 49
10. $x^2 + y^3$ for $x = 5$ and $y = -3$ — −2
11. $a^b + 6^c$ for $a = 3$, $b = 4$, and $c = 2$ — 117
12. $s^4 + (t - r)^5$ for $s = 2$, $t = 7$, and $r = 8$ — 15
13. $y^x + x^y - xy$ for $x = 2$ and $y = 3$ — 11
14. $5x^9 - (y + z)$ for $x = -1$, $y = 10$, and $z = -6$ — −9
15. $x^z \div y^x$ for $x = 2$, $y = 4$, and $z = 5$ — 2
16. $ac - b^c$ for $a = 12$, $b = -6$, and $c = 2$ — −12

17. Find the area of a circle if the radius is 20 in. (Hint: $A = \pi r^2$.) Use 3.14 for π.
$1256\ in^2$

18. If the area of a regular pentagon is $A = 1.720a^2$, in which a is one of the sides. Find the area of a regular pentagon with a side that measures 50 cm.
$4300\ cm^2$

LESSON 4-1
Reteach
Exponents

The fifth power of 3 $3^5 = 3 \cdot 3 \cdot 3 \cdot 3 \cdot 3$ (base: 3, exponent: 5; 3 used as a factor 5 times)

Complete to write each expression using an exponent. State the power.

2 1. $5 \cdot 5 \cdot 5 \cdot 5 = 5^{4}$ the fourth power of 5
3 2. $(-7) \cdot (-7) \cdot (-7) = (-7)^{3}$ the third power of −7

Complete to evaluate each expression.

1 3. $(-2)^3 = (-2)(-2)(-2) =$ −8
5 4. $10^4 =$ 10 • 10 • 10 • 10 = 10,000
5 5. $(-5)^4 = (-5)(-5)(-5)(-5) =$ 625

When an expression is a product that includes a power, you simplify the power first.
$3 \cdot 2^3 = 3 \cdot 2 \cdot 2 \cdot 2 = 3 \cdot 8 = 24$

Complete to simplify each expression.

4 6. $4 \cdot (-2)^3 = 4(-2)(-2)(-2) =$ −32
5 7. $5 \cdot 3^3 =$ 5 • 3 • 3 • 3 = 135
4 8. $(3 \cdot 2)^3 = 6^3 =$ 6 • 6 • 6 = 216
5 9. $(-4(-2))^3 = (8)^3 = (8)(8)(8) =$ 512

4 10. 25 − 3(4 • 3^2)
= 25 − 3(4 • 9)
= 25 − 3(36)
= 25 − 108
= −83

4 11. −100 − 2(3 • $4)^2$
= −100 − 2($12)^2$
= −100 − 2(144)
= −100 − 288
= −388

4 12. 15 − 4(3 + 3^2)
= 15 − 4(3 + 9)
= 15 − 4(12)
= 15 − 48
= −33

/40

LESSON 4-1 Challenge
Check This Out

Imagine an 8 × 8 checkerboard.

- Put 1 penny on the first square.
- Stack 2 pennies on an adjacent square.
- Stack 4 pennies on an adjacent square.
- Stack 8 pennies on an adjacent square.

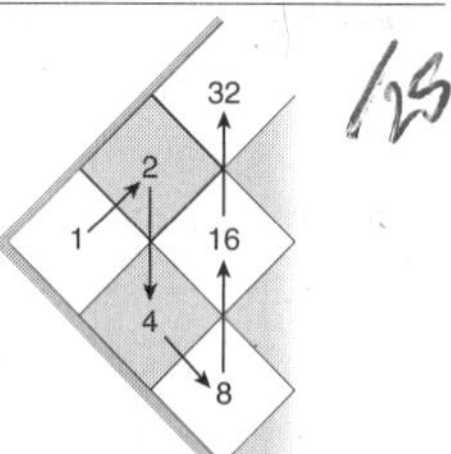

Assume the pattern continues so that each square has double the number of pennies as the previous square.

Complete the table.

	Square	1	2	3	4	5	6	7	8
1.	Number of Pennies	1	2	4	8	16	32	64	128
2.	Exponent Form	2^0	2^1	2^2	2^3	2^4	2^5	2^6	2^7

3. Look for a pattern in the table. How many pennies are on the 12th square? the 15th square?

   square $n = 2^{n-1}$ pennies; square 12 = 2^{11} or 2048 pennies;
   square 15 = 2^{14} or 16,384 pennies

4. Find the number of pennies on the 25th square. How much money is this? Estimate the height of the stack.

   2^{24} or 16,777,216 pennies = $167,772.16;
   about 1 million in., or over 15 mi

5. Find the number of pennies on the 10th square. Find the total number of pennies on the first nine squares. Which number is greater?

   512 pennies on 10th square; 511 pennies on the first nine squares.
   $512 > 511$

6. Explain why it is impossible to stack pennies on every square in the manner described at the top of the page.

   Possible answer: By the 25th square, the stack is already over 15 mi high, and there are 39 more squares to stack.

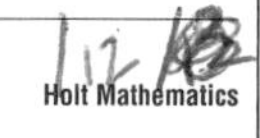

LESSON 4-1 Problem Solving
Exponents

Write the correct answer.

1. The formula for the volume of a cube is $V = e^3$ where e is the length of a side of the cube. Find the volume of a cube with side length 6 cm.

   216 cm^3

2. The distance in feet traveled by a falling object is given by the formula $d = 16t^2$ where t is the time in seconds. Find the distance an object falls in 4 seconds.

   256 feet

3. The surface area of a cube can be found using the formula $S = 6e^2$ where e is the length of a side of the cube. Find the surface area of a cube with side length 6 cm.

   216 cm^2

4. John's father offers to pay him 1 cent for doing the dishes the first night, 2 cents for doing the dishes the second, 4 cents for the third, and so on, doubling each night. Write an expression using exponents for the amount John will get paid on the tenth night.

   2^9 cents

Use the table below for Exercises 5–7, which shows the number of e-mails forwarded at each level if each person continues a chain by forwarding an e-mail to 10 friends. Choose the letter for the best answer.

Forwarded E-mails

Level	E-mails forwarded
1	10
2	100
3	1000
4	10,000

5. How many e-mails were forwarded at level 5 alone?

   A 5^{10} C 2^{10}
   B 2^5 (D) 10^5

6. How many e-mails were forwarded at level 6 alone?

   F 100,000 H 10,000,000
   (G) 1,000,000 J 100,000,000

7. Forwarding chain e-mails can create problems for e-mail servers. Find out how many total e-mails have been forwarded after 6 levels.

   (A) 1,111,110 C 1,000,000
   B 6,000,000 D 100,000,000

LESSON 4-1 Reading Strategies
Multiple Meanings

Exponents are an efficient way to express repeated multiplication.

Example: 4^5 → is read "4 to the fifth power."
4^5 means **4** is a factor **5** times: 4 × 4 × 4 × 4 × 4.
$4^5 = 1024$ → is read "4 to the fifth power equals 1024," or "the value of 4 to the fifth power is 1024."

The **base** identifies the factor.
The **exponent** identifies how many times the base is a factor.

base → 4^5 ← exponent

A negative number with an even exponent will be positive. → $(-8)^4 = (-8)(-8)(-8)(-8)$
$(-8)^4 = 4096$
"negative 8 to the fourth power"

A negative number with an odd exponent will be negative. → $(-8)^3 = (-8)(-8)(-8)$
$(-8)^3 = -512$
"negative 8 to the third power"

Answer each question.

1. How do you read 7^4?

   seven to the fourth power

2. What does 7^4 mean?

   $7 \times 7 \times 7 \times 7$

3. What is the value of 7^4? 2401

4. How do you read $(-3)^5$?

   negative 3 to the fifth power

5. What does $(-3)^5$ mean?

   $(-3)(-3)(-3)(-3)(-3)$

6. How you can tell when a negative number raised to a power will have a negative value?

   If the exponent is an odd number, the value will be negative.

7. Will the value of $(-6)^2$ be positive or negative? positive

8. Will the value of $(-6)^5$ be positive or negative? negative

LESSON 4-1 Puzzles, Twisters & Teasers
Something's Fishy!

Rewrite each of the following using exponents. Then solve the riddle.

H	4	4^1
Y	8 • 8	8^2
S	(−9) • (−9) • (−9)	$(-9)^3$
O	(−3) • (−3) • (−3) • (−3) • (−3) • (−3)	$(-3)^6$
I	6 • 6 • 6 • 6 • 6 • 6	6^6
U	7 • 7 • 7 • 7	7^4
F	2 • 2 • 2 • 2 • 2 • 2 • 2 • 2 • 2	2^9
C	(−6) • (−6) • (−6) • (−6) • (−6)	$(-6)^5$
A	8 • 8 • 8 • 8	8^4
N	(−12) • (−12) • (−12)	$(-12)^3$
T	7 • 7 • 7	7^3
X	(−5) • (−5)	$(-5)^2$

What's the difference between a guitar and a fish?

Y 8^2 O $(-3)^6$ U 7^4

C $(-6)^5$ A 8^4 N $(-12)^3$ ' T 7^3

T 7^3 U 7^4 N $(-12)^3$ A 8^4

F 2^9 I 6^6 S $(-9)^3$ H 4^1.

LESSON **4-2**

Practice A

Look for a Pattern in Integer Exponents

Evaluate the powers of 10.

1. 10^{-1} 0.1
2. 10^{-6} $\frac{1}{1,000,000}$
3. 10^{2} 100
4. 10^{1} 10
5. 10^{0} 1
6. 10^{3} 1000
7. 10^{-5} 0.00001
8. 10^{6} 1,000,000
9. 10^{-7} $\frac{1}{10,000,000}$
10. 10^{4} 10,000
11. 10^{-3} 0.001
12. 10^{5} 100,000

Evaluate.

13. $(-2)^{-3}$ $-\frac{1}{8}$
14. 3^{-4} $\frac{1}{81}$
15. $(-4)^{-2}$ $\frac{1}{16}$
16. 2^{-4} $\frac{1}{16}$
17. 5^{-2} $\frac{1}{25}$
18. 6^{-3} $\frac{1}{216}$
19. $(-9)^{-2}$ $\frac{1}{81}$
20. $(-3)^{-3}$ $-\frac{1}{27}$
21. $8 - 3^{0} + 2^{-1}$ $7\frac{1}{2}$
22. $4 + (-6)^{0} - 4^{-1}$ $4\frac{3}{4}$
23. $3(-9)^{0} + 4^{-2}$ $3\frac{1}{16}$
24. $6 + (-5)^{-2} - (4 + 3)^{0}$ $5\frac{1}{25}$
25. One centimeter equals 10^{-2} meter. Evaluate 10^{-2}.
 $\frac{1}{100}$
26. The area of a square is 10^{-4} square teet. Evaluate 10^{-4}.
 $\frac{1}{10,000}$

LESSON **4-2**

Practice B

Look for a Pattern in Integer Exponents

Evaluate the powers of 10.

1. 10^{-3} $\frac{1}{1000}$ 0.001
2. 10^{3} 1000
3. 10^{-5} $\frac{1}{100000}$ 0.00001
4. 10^{-2} $\frac{1}{100}$ 0.01
5. 10^{0} 1
6. 10^{4} 10,000
7. 10^{1} 10
8. 10^{5} 100,000

Evaluate.

9. $(-6)^{-2}$ $\frac{1}{36}$
10. $(-9)^{-3}$ $-\frac{1}{729}$
11. 2^{-5} $\frac{1}{32}$
12. $(-3)^{-4}$ $\frac{1}{81}$
13. $(-12)^{-1}$ $-\frac{1}{12}$
14. 6^{-3} $\frac{1}{216}$
15. $10 - (3 + 2)^{0} + 2^{-1}$ $9\frac{1}{2}$
16. $15 + (-6)^{0} - 3^{-2}$ $15\frac{8}{9}$
17. $6(8 - 2)^{0} + 4^{-2}$ $6\frac{1}{16}$
18. $2^{-2} + (-4)^{-1}$ 0
19. $3(1 - 4)^{-2} + 9^{-1} + 12^{0}$ $1\frac{4}{9}$
20. $9^{0} + 64(3 + 5)^{-2}$ 2
21. One milliliter equals 10^{-3} liter. Evaluate 10^{-3}.
 $\frac{1}{1000}$ 0.001
22. The volume of a cube is 10^{6} cubic feet. Evaluate 10^{6}.
 1,000,000 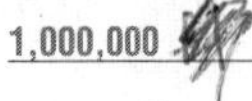

LESSON **4-2**

Practice C

Look for Patterns in Integer Exponents

Evaluate.

1. $(-4)^{-3}$ $-\frac{1}{64}$
2. 11^{-2} $\frac{1}{121}$
3. 9^{-3} $\frac{1}{729}$
4. 3^{-5} $\frac{1}{243}$
5. $(-2)^{-6}$ $\frac{1}{64}$
6. $(-4)^{-5}$ $-\frac{1}{1024}$
7. $4^{-1} - (3 + 2)^{0} + 2^{-2}$ $-\frac{1}{2}$
8. $7 - (-6)^{0} - 3^{-2}$ $5\frac{8}{9}$
9. $3^{2}(8 + 6)^{0} + 4^{-2}$ $9\frac{1}{16}$
10. $2^{-2} - 8 + (-4)^{-1}$ -8
11. $7(-3 - 4)^{-2} + 2^{-1}$ $\frac{9}{14}$
12. $(9 - 3)^{0} + 64(9 - 5)^{-2}$ 5
13. $5^{2} + (-5)^{0} + (4^{-3} + 2^{-6})$ $26\frac{1}{32}$
14. $108(4 + 2)^{-3} + (10^{0} - 5)^{2}$ $16\frac{1}{2}$
15. Super Bowl XXXVI was held in New Orleans, Louisiana. The New England Patriots played the St. Louis Rams. The St. Louis fans traveled about 1127 km to New Orleans. How many meters is this? (Hint: 1 km = 10^{3} m.)
 1,127,000 m
16. The New England fans traveled 2,420,453 m. How many kilometers is this?
 2420 km

/16

LESSON **4-2**

Reteach

Look for a Pattern in Integer Exponents

To rewrite a negative exponent, move the power to the denominator of a unit fraction. $5^{-2} = \frac{1}{5^2}$

/62

Complete to rewrite each power with a positive exponent.

1. $7^{-3} = \frac{1}{7^3}$
2. $9^{-5} = \frac{1}{9^5}$
3. $13^{-4} = \frac{1}{13^4}$

Complete each pattern.

4. $10^{-1} = \frac{1}{10} = 0.1$
 $10^{-2} = \frac{1}{10^2} = \frac{1}{100} = 0.01$
 $10^{-3} = \frac{1}{10^3} = \frac{1}{1000} = 0.001$
5. $5^{-1} = \frac{1}{5}$
 $5^{-2} = \frac{1}{5^2} = \frac{1}{5 \cdot 5} = \frac{1}{25}$
 $5^{-3} = \frac{1}{5^3} = \frac{1}{5 \cdot 5 \cdot 5} = \frac{1}{125}$
6. $3^{-1} = \frac{1}{3}$
 $3^{-2} = \frac{1}{3^2} = \frac{1}{3 \cdot 3} = \frac{1}{9}$
 $3^{-3} = \frac{1}{3^3} = \frac{1}{3 \cdot 3 \cdot 3} = \frac{1}{27}$
7. $(-4)^{-1} = \frac{1}{-4}$
 $(-4)^{-2} = \frac{1}{(-4)^2} = \frac{1}{(-4) \cdot (-4)} = \frac{1}{16}$
 $(-4)^{-3} = \frac{1}{(-4)^3} = \frac{1}{(-4) \cdot (-4) \cdot (-4)} = -\frac{1}{64}$

Evaluate.

8. $2^{-3} = \frac{1}{} =$ 2^3; $\frac{1}{8}$
9. $(-6)^{-2} = \frac{1}{} =$ $(-6)^2$; $\frac{1}{36}$
10. $4^{-2} = \frac{1}{} =$ 4^2; $\frac{1}{16}$
11. $(-3)^{-3} = \frac{1}{} =$ $(-3)^3$; $-\frac{1}{27}$
12. $6^{-2} =$ $\frac{1}{36}$
13. $(-2)^{-3} =$ $-\frac{1}{8}$
14. $6^{-3} =$ $\frac{1}{216}$
15. $(-5)^{-2} =$ $\frac{1}{25}$
16. $2^{-4} =$ $\frac{1}{16}$
17. $(-9)^{-1} =$ $-\frac{1}{9}$

LESSON 4-2
Challenge
Stuff It!

$9^{\frac{1}{2}}$ means $\sqrt[2]{9^1}$.

To find the value, first evaluate the root: $\sqrt[2]{9} = 3$.

Then, raise the result to the indicated power: $3^1 = 3$.

So, $9^{\frac{1}{2}}$ is $\sqrt[2]{9^1} = 3^1 = 3$.

In general, here's the way to rewrite a term with a fractional exponent:

$$x^{\frac{a}{b}} = \sqrt[b]{x^a}$$

Evaluate $8^{\frac{2}{3}}$.

$8^{\frac{2}{3}} = \sqrt[3]{8^2}$ Rewrite using radical form.

$= 2^2$ Evaluate the root; $\sqrt[3]{8} = 2$ since $2 \cdot 2 \cdot 2 = 8$.

$= 4$ Evaluate the power.

Rewrite each term using radical form. Evaluate the root. Evaluate the power.

1. $64^{\frac{1}{2}} = \sqrt[2]{64^1} = 8^1 = 8$
2. $100^{\frac{1}{2}} = \sqrt[2]{100^1} = 10^1 = 10$
3. $400^{\frac{1}{2}} = \sqrt[2]{400^1} = 20^1 = 20$
4. $64^{\frac{2}{3}} = \sqrt[3]{64^2} = 4^2 = 16$
5. $216^{\frac{2}{3}} = \sqrt[3]{216^2} = 6^2 = 36$
6. $1000^{\frac{2}{3}} = \sqrt[3]{1000^2} = 10^2 = 100$
7. $625^{\frac{3}{4}} = \sqrt[4]{625^3} = 5^3 = 125$
8. $32^{\frac{2}{5}} = \sqrt[5]{32^2} = 2^2 = 4$
9. $10{,}000^{\frac{5}{4}} = \sqrt[4]{10{,}000^5} = 10^5 = 100{,}000$

LESSON 4-2
Problem Solving
Look for a Pattern in Integer Exponents

Write the correct answer.

1. The weight of 10^7 dust particles is 1 gram. Evaluate 10^7.
   10,000,000
2. The weight of one dust particle is 10^{-7} gram. Evaluate 10^{-7}.
   0.0000001
3. As of 2001, only 10^6 rural homes in the United States had broadband Internet access. Evaluate 10^6.
   1,000,000
4. Atomic clocks measure time in microseconds. A microsecond is 10^{-6} second. Evaluate 10^{-6}.
   0.000001

Choose the letter for the best answer.

5. The diameter of the nucleus of an atom is about 10^{-15} meter. Evaluate 10^{-15}.
   A 0.0000000000001
   B 0.00000000000001
   C 0.0000000000000001
   (D) 0.000000000000001
6. The diameter of the nucleus of an atom is 0.000001 nanometer. How many nanometers is the diameter of the nucleus of an atom?
   F $(-10)^5$
   G $(-10)^6$
   (H) 10^{-6}
   J 10^{-5}
7. A ruby-throated hummingbird weighs about 3^{-2} ounce. Evaluate 3^{-2}.
   A -9
   B -6
   (C) $\frac{1}{9}$
   D $\frac{1}{6}$
8. A ruby-throated hummingbird breathes 2×5^3 times per minute while at rest. Evaluate this amount.
   F 1,000
   (G) 250
   H 125
   J 30

LESSON 4-2
Reading Strategies
Using Patterns

The pattern in this table will help you evaluate powers with exponents.

Look at the pattern of the products in the first column. You see that as you move down the column the products are getting smaller. That is because there is one less factor. Each product is divided by 2 to get the next product.

Column 1	Column 2	Column 3
$2^3 = 8$	$3^3 = 27$	$4^3 = 64$
$2^2 = 4$	$3^2 = 9$	$4^2 = 16$
$2^1 = 2$	$3^1 = 3$	$4^1 = 4$
$2^0 = 1$	$3^0 = 1$	$4^0 = 1$
$2^{-1} = \frac{1}{2}$	$3^{-1} = \frac{1}{3}$	$4^{-1} = \frac{1}{4}$
$2^{-2} = \frac{1}{4}$	$3^{-2} = \frac{1}{9}$	$4^{-2} = \frac{1}{16}$

Look at the second and third columns to answer Exercises 1-6.

1. What is the base in Column 2? 3
2. What is the product divided by each time to get the next product? 3
3. What is $1 \div 3$? $\frac{1}{3}$
4. What is the base in Column 3? 4
5. What number is the product divided by each time to get the next product? 4
6. What is $\frac{1}{4} \div 4$? $\frac{1}{16}$

Complete the table using the table above as a guide.

Column 1	Column 2	Column 3
$5^3 = 125$	$6^3 = 216$	$10^3 = 1000$
$5^2 = 25$	$6^2 = 36$	$10^2 = 100$
$5^1 = 5$	$6^1 = 6$	$10^1 = 10$
$5^0 = 1$	$6^0 = 1$	$10^0 = 1$
$5^{-1} = \frac{1}{5}$	$6^{-1} = \frac{1}{6}$	$10^{-1} = \frac{1}{10}$
$5^{-2} = \frac{1}{25}$	$6^{-2} = \frac{1}{36}$	$10^{-2} = \frac{1}{100}$

LESSON 4-2
Puzzles, Twisters & Teasers
An Alarming Activity!

Decide whether or not each equation is correct. Circle the letters above your answers. Then solve the riddle.

	correct	incorrect
1. $(-2)^{-4} = 0.0002$	O	(A)
2. $10^{-5} = 0.00001$	(L)	Z
3. $10^{-2} = 1.0$	F	(A)
4. $2^{-3} = 0.3$	D	(R)
5. $10^{-3} = 0.001$	(M)	S
6. $(-4)^{-3} = 4.004$	Q	(C)
7. $10^{-4} = 0.0001$	(L)	W
8. $10^2 = 100$	(U)	E
9. $3^{-2} = 30.0$	T	(C)
10. $10^0 = 1$	(K)	P

What do you call a rooster that wakes you up crowing?

An A L A R M C L U C K

LESSON 4-3 Practice A
Properties of Exponents

Multiply. Write the product as one power.

1. $2^2 \cdot 2^3$ $\quad$ 2^5
2. $3^5 \cdot 3^2$ $\quad$ 3^7
3. $1^3 \cdot 1^5$ $\quad$ 1^8
4. $5^4 \cdot 5^3$ $\quad$ 5^7
5. $8^1 \cdot 8^1$ $\quad$ 8^2
6. $7^4 \cdot 7^5$ $\quad$ 7^9
7. $12^1 \cdot 12^2$ $\quad$ 12^3
8. $n^3 \cdot n^8$ $\quad$ n^{11}

Divide. Write the quotient as one power.

9. $\frac{2^5}{2^2}$ $\quad$ 2^3
10. $\frac{10^4}{10^3}$ $\quad$ 10^1
11. $\frac{4^6}{4^3}$ $\quad$ 4^3
12. $\frac{(-3)^6}{(-3)^4}$ $\quad$ $(-3)^2$
13. $\frac{5^8}{5^6}$ $\quad$ 5^2
14. $\frac{24^9}{24^3}$ $\quad$ 24^6
15. $\frac{(-6)^8}{(-6)^5}$ $\quad$ $(-6)^3$
16. $\frac{b^7}{b^5}$ $\quad$ b^2

Simplify.

17. $(3^2)^4$ $\quad$ 3^8
18. $(6^3)^{-1}$ $\quad$ 6^{-3}
19. $(4^5)^0$ $\quad$ 4^0
20. $(8^2)^3$ $\quad$ 8^6
21. $(5^{-2})^3$ $\quad$ 5^{-6}
22. $(7^0)^4$ $\quad$ 7^0
23. $(9^4)^{-2}$ $\quad$ 9^{-8}
24. $(s^5)^2$ $\quad$ s^{10}

25. The Haywood Paper Company has 5^2 warehouses. Each warehouse holds 5^5 boxes of paper. How many boxes of paper are stored in all the warehouses? Write the answer as one power.

5^7

26. Write the expression for 5 used as a factor eight times being divided by 5 used as a factor six times. Simplify the expression as one power.

$\frac{5^8}{5^6} = 5^2$

LESSON 4-3 Practice B
Properties of Exponents

Multiply. Write the product as one power.

1. $10^5 \cdot 10^7$ $\quad$ 10^{12}
2. $x^9 \cdot x^8$ $\quad$ x^{17}
3. $14^7 \cdot 14^9$ $\quad$ 14^{16}
4. $12^6 \cdot 12^8$ $\quad$ 12^{14}
5. $y^{12} \cdot y^{10}$ $\quad$ y^{22}
6. $15^9 \cdot 15^{14}$ $\quad$ 15^{23}
7. $(-11)^{20} \cdot (-11)^{10}$ $\quad$ $(-11)^{30}$
8. $(-a)^6 \cdot (-a)^7$ $\quad$ $(-a)^{13}$

Divide. Write the quotient as one power.

9. $\frac{12^9}{12^2}$ $\quad$ 12^7
10. $\frac{(-11)^{12}}{(-11)^8}$ $\quad$ $(-11)^4$
11. $\frac{x^{10}}{x^5}$ $\quad$ x^5
12. $\frac{16^{10}}{16^2}$ $\quad$ 16^8
13. $\frac{17^{19}}{17^2}$ $\quad$ 17^{17}
14. $\frac{14^{15}}{14^{13}}$ $\quad$ 14^2
15. $\frac{23^{17}}{23^9}$ $\quad$ 23^8
16. $\frac{(-a)^{12}}{(-a)^7}$ $\quad$ $(-a)^5$

Simplify.

17. $(6^2)^4$ $\quad$ 6^8
18. $(2^4)^{-3}$ $\quad$ 2^{-12}
19. $(3^5)^{-1}$ $\quad$ 3^{-5}
20. $(y^5)^2$ $\quad$ y^{10}
21. $(9^{-2})^3$ $\quad$ 9^{-6}
22. $(10^0)^3$ $\quad$ 10^0
23. $(x^4)^{-2}$ $\quad$ x^{-8}
24. $(5^{-2})^0$ $\quad$ 5^0

Write the product or quotient as one power.

25. $\frac{w^{12}}{w^3}$ $\quad$ w^9
26. $d^8 \cdot d^5$ $\quad$ d^{13}
27. $(-15)^5 \cdot (-15)^{10}$ $\quad$ $(-15)^{15}$

28. Jefferson High School has a student body of 6^4 students. Each class has approximately 6^2 students. How many classes does the school have? Write the answer as one power.

6^2

29. Write the expression for a number used as a factor fifteen times being multiplied by a number used as a factor ten times. Then, write the product as one power.

$x^{15} \cdot x^{10} = x^{25}$

LESSON 4-3 Practice C
Properties of Exponents

Write the product or quotient as one power.

1. $x^{10} \cdot x^8$ $\quad$ x^{18}
2. $(-10)^{14} \cdot (-10)^4$ $\quad$ $(-10)^{18}$
3. $\frac{d^5}{d}$ $\quad$ d^4
4. $9^{10} \div 9^2$ $\quad$ 9^8
5. $t^{12} \cdot t^5$ $\quad$ t^{17}
6. $\frac{(-x)^{10}}{(-x)^8}$ $\quad$ $(-x)^2$
7. $16^8 \div 16^8$ $\quad$ 16^0
8. $14^9 \cdot 14^9$ $\quad$ 14^{18}
9. $(-k)^{12} \div (-k)^9$ $\quad$ $(-k)^3$
10. $15 \cdot 15^{11}$ $\quad$ 15^{12}
11. $17^{10} \cdot 17$ $\quad$ 17^{11}
12. $x^8 \div x^8$ $\quad$ x^0

Simplify.

13. $(5^3)^4$ $\quad$ 5^{12}
14. $(7^4)^{-2}$ $\quad$ 7^{-8}
15. $(2^6)^{-1}$ $\quad$ 2^{-6}
16. $(x^2)^5$ $\quad$ x^{10}
17. $(3^{-2})^{-3}$ $\quad$ 3^6
18. $(12^0)^{-2}$ $\quad$ 12^0
19. $(w^4)^2$ $\quad$ w^8
20. $(y^{-1})^6$ $\quad$ y^{-6}

Write as one power.

21. $x^3 \cdot x^2 \cdot x^4$ $\quad$ x^9
22. $5^5 \div 5^2 \cdot 5^3$ $\quad$ 5^6
23. $8^4 \cdot 8 \div 8^2$ $\quad$ 8^3
24. $x^7 \div x^2 \cdot x^3$ $\quad$ x^2
25. $(-4)^4 \cdot (-4)^6 \div (-4)^2$ $\quad$ $(-4)^8$
26. $2^5 \div 2 \cdot 2^2 \div 2^3$ $\quad$ 2^3

27. Justine and ninety-nine of her co-workers won the lottery worth $\$10^8$. They all received their winnings over ten years. How much did each receive in a one-year period?

$\$10^5$ or \$100,000

28. A number to the 7th power divided by the same number to the 3rd power equals 256. What is the number?

4

LESSON 4-3 Reteach
Properties of Exponents

To multiply powers with the same base, keep the base and add exponents.	To divide powers with the same base, keep the base and subtract exponents.	To raise a power to a power, keep the base and multiply exponents.
$x^a \cdot x^b = x^{a+b}$	$x^a \div x^b = x^{a-b}$	$(x^a)^b = x^{ab}$
$4^5 \cdot 4^2 = 4^{5+2} = 4^7$	$4^5 \div 4^2 = 4^{5-2} = 4^3$	$(4^5)^2 = 4^{5(2)} = 4^{10}$
$8^3 \cdot 8 = 8^{3+1} = 8^4$	$8^3 \div 8 = 8^{3-1} = 8^2$	

Complete to see why the rules for exponents work.

1. $4^5 \cdot 4^2 = (4)(4)(4)(4)(4) \cdot (4)(4) = 4^7$
2. $8^3 \cdot 8 = (8)(8)(8) \cdot (8) = 8^4$
3. $4^5 \div 4^2 = \frac{4^5}{4^2} = \frac{4 \cdot 4 \cdot 4 \cdot 4 \cdot 4}{4 \cdot 4} = 4^3$
4. $8^3 \div 8 = \frac{8^3}{8} = \frac{8 \cdot 8 \cdot 8}{8} = 8^2$
5. $(4^2)^3 = 4^2 \cdot 4^2 \cdot 4^2 = 4^{2+2+2} = 4^{2(3)} = 4^6$

Complete to write each product or quotient as one power.

6. $12^3 \cdot 12^2 = 12^{3+2} = 12^5$
7. $9^4 \cdot 9^3 = 9^{4+3} = 9^7$
8. $\frac{7^6}{7^2} = 7^{6-2} = 7^4$
9. $\frac{12^6}{12^4} = 12^{6-4} = 12^2$

Write each product or quotient as one power.

10. $10^4 \cdot 10^6 = 10^{10}$
11. $5^5 \cdot 5 = 5^6$
12. $4^5 \cdot 4 \cdot 4^3 = 4^9$
13. $\frac{15^6}{15^2} = 15^4$
14. $\frac{9^5}{9} = 9^4$
15. $\frac{2^{10}}{2^2} = 2^8$

Simplify.

16. $(5^3)^4 = 5^{3(4)} = 5^{12}$
17. $(6^2)^4 = 6^{2(4)} = 6^8$
18. $(2^5)^2 = 2^{10}$

LESSON 4-3
Challenge
Square Dance

Study these patterns.

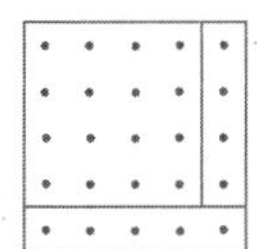

$1 = 1^2$

$1^2 + 1 + 2 = 4 = 2^2$

$2^2 + 2 + 3 = 9 = 3^2$

$3^2 + 3 + 4 = 16 = 4^2$

So, according to the pattern, 5^2 can be written as the sum of 4^2 and two consecutive integers.

1. Draw a diagram and write an equation to illustrate 5^2.

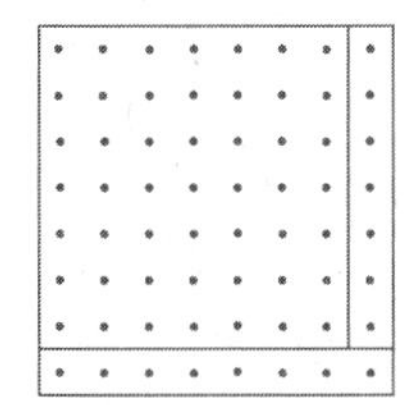

Equation: $4^2 + 4 + 5 = 25 = 5^2$

2. Draw a diagram and write an equation to illustrate 8^2.

Equation: $7^2 + 7 + 8 = 64 = 8^2$

3. Use the pattern to write an equation to indicate that, for any integer n, $(n + 1)^2$ can be written as the sum of n^2 and two consecutive integers.

Equation: $n^2 + (n) + (n + 1) = (n + 1)^2$

4. If you know that $20^2 = 400$, use the pattern to calculate 21^2.

$21^2 =$ $400 + 20 + 21 = 441$

LESSON 4-3
Problem Solving
Properties of Exponents

Write each answer as a power.

1. Cindy separated her fruit flies into equal groups. She estimates that there are 2^{10} fruit flies in each of 2^2 jars. How many fruit flies does Cindy have in all?

2^{12} fruit flies

2. Suppose a researcher tests a new method of pasteurization on a strain of bacteria in his laboratory. If the bacteria are killed at a rate of 8^9 per sec, how many bacteria would be killed after 8^2 sec?

8^{11} bacteria

3. A satellite orbits the earth at about 13^4 km per hour. How long would it take to complete 24 orbits, which is a distance of about 13^5 km?

13 hr

4. The side of a cube is 3^4 centimeters long. What is the volume of the cube? (Hint: $V = s^3$.)

3^{12} cm

Use the table to answer Exercises 5–6. The table describes the number of people involved at each level of a pyramid scheme. In a pyramid scheme each individual recruits so many others to participate who in turn recruit others, and so on. Choose the letter of the best answer.

5. Using exponents, how many people will be involved at level 6?

A 6^6 C 5^5
B 6^5 (D) 5^6

6. How many times more people will be involved at level 6 than at level 2?

(F) 5^4 H 5^5
G 5^3 J 5^6

Pyramid Scheme

Each person recruits 5 others.

Level	Total Number of People
1	5
2	5^2
3	5^3
4	5^4

7. There are 10^3 ways to make a 3-digit combination, but there are 10^6 ways to make a 6-digit combination. How many times more ways are there to make a 6-digit combination than a 3-digit combination?

A 5^{10} C 2^5
B 2^{10} (D) 10^3

8. After 3 hours, a bacteria colony has $(25^3)^3$ bacteria present. How many bacteria are in the colony?

F 25^1 (H) 25^9
G 25^6 J 25^{33}

LESSON 4-3
Reading Strategies
Organization Patterns

There are some rules that make multiplying or dividing exponents with the same base easier.

To multiply powers with the same base, add exponents.

$(4 \cdot 4) \cdot (4 \cdot 4 \cdot 4) = 4 \cdot 4 \cdot 4 \cdot 4 \cdot 4$

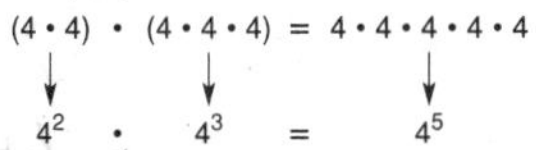

$4^2 \cdot 4^3 = 4^5$

The base of 4 is the same, so: $4^2 \cdot 4^3 = 4^{2+3} = 4^5$.

To divide powers with the same base, subtract the exponents.

$\frac{6 \cdot 6 \cdot 6 \cdot 6 \cdot 6 \cdot 6}{6 \cdot 6 \cdot 6} = \frac{6^6}{6^3}$

$\frac{6 \cdot 6 \cdot 6 \cdot \not{6} \cdot \not{6} \cdot \not{6}}{\not{6} \cdot \not{6} \cdot \not{6}} = 6^3$

The base of 6 is the same, so: $\frac{6^6}{6^3} = 6^{6-3} = 6^3$

Answer each question.

1. What is the base for 3^2? 3
2. What is the base for 3^4? 3
3. Are the bases the same for these powers? yes
4. Write all the factors for $3^2 \cdot 3^4$. $(3 \cdot 3) \cdot (3 \cdot 3 \cdot 3 \cdot 3)$
5. Add the exponents for 3^2 and 3^4 and rewrite the number using the same base. 3^6
6. Are the answers for Exercise 4 and Exercise 5 the same? yes
7. Are the bases the same for $5^5 \div 5^2$? yes
8. Subtract the exponents and rewrite the problem. 5^3

LESSON 4-3
Puzzles, Twisters & Teasers
Button It Up!

Find and circle words from the list in the word search. Then find a word that answers the riddle. Circle it and write it on the line.

factor, power, property, exponent, group
base, combine, evaluate, zero, apply

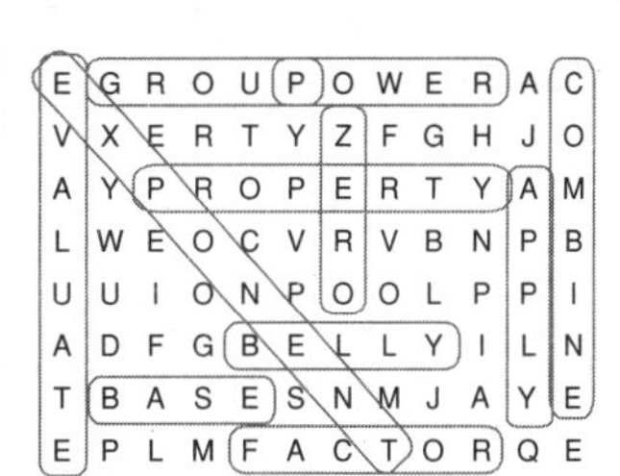

E	G	R	O	U	P	O	W	E	R	A	C
V	X	E	R	T	Y	Z	F	G	H	J	O
A	Y	P	R	O	P	E	R	T	Y	A	M
L	W	E	O	C	V	R	V	B	N	P	B
U	U	I	O	N	P	O	O	L	P	P	I
A	D	F	G	B	E	L	L	Y	I	L	N
T	B	A	S	E	S	N	M	J	A	Y	E
E	P	L	M	F	A	C	T	O	R	Q	E

What kind of button can't you unbutton?

a B E L L Y **button**

LESSON 4-4

Practice A

Scientific Notation

Write each number in standard notation.

1. 1.76×10^{-1} — 0.176
2. 8.9×10^{3} — 8,900
3. 6.2×10^{-2} — 0.062
4. 1.01×10^{2} — 101
5. 5.8×10^{4} — 58,000
6. 8.1×10^{5} — 810,000
7. 3.8×10^{-4} — 0.00038
8. 2.03×10^{-3} — 0.00203
9. 5.0×10^{3} — 5000
10. 3.12×10^{5} — 312,000
11. 7.6×10^{-2} — 0.076
12. 8.54×10^{-5} — 0.0000854

Write each number in scientific notation.

13. 376,000 — 3.76×10^{5}
14. 9,580,000 — 9.58×10^{6}
15. 650 — 6.5×10^{2}
16. 1006 — 1.006×10^{3}
17. 29 — 2.9×10^{1}
18. 0.0061 — 6.1×10^{-3}
19. 0.0107 — 1.07×10^{-2}
20. 0.0002008 — 2.008×10^{-4}
21. 0.00053 — 5.3×10^{-4}
22. 250,800 — 2.508×10^{5}
23. 0.000094 — 9.4×10^{-5}
24. 0.00086 — 8.6×10^{-4}

25. Earth is about 93,000,000 miles from the Sun. Write this number in scientific notation.

9.3×10^{7}

26. The diameter of Earth is about 1.276×10^{4} kilometers. The diameter of Venus is about 1.21×10^{4} kilometers. Which planet has the greater diameter, Earth or Venus?

Earth

LESSON 4-4

Practice B

Scientific Notation

Write each number in standard notation.

1. 2.54×10^{2} — 254
2. 6.7×10^{-2} — 0.067
3. 1.14×10^{3} — 1140
4. 3.8×10^{-1} — 0.38
5. 7.53×10^{-3} — 0.00753
6. 5.6×10^{4} — 56,000
7. 9.1×10^{5} — 910,000
8. 6.08×10^{-4} — 0.000608
9. 8.59×10^{5} — 859,000
10. 3.331×10^{6} — 3,331,000
11. 7.21×10^{-3} — 0.00721
12. 5.88×10^{-4} — 0.000588

Write each number in scientific notation.

13. 75,000,000 — 7.5×10^{7}
14. 208 — 2.08×10^{2}
15. 907,100 — 9.071×10^{5}
16. 56 — 5.6×10^{1}
17. 0.093 — 9.3×10^{-2}
18. 0.00006 — 6.0×10^{-5}
19. 0.00852 — 8.52×10^{-3}
20. 0.0505 — 5.05×10^{-2}
21. 0.003007 — 3.007×10^{-3}
22. 5226 — 5.226×10^{3}
23. 0.04 — 4.0×10^{-2}
24. 98,856 — 9.8856×10^{4}

25. Jupiter is about 778,120,000 kilometers from the Sun. Write this number in scientific notation.

7.7812×10^{8}

26. The *E. coli* bacterium is about 5×10^{-7} meters wide. A hair is about 1.7×10^{-5} meters wide. Which is wider, the bacterium or the hair?

the hair

LESSON 4-4

Practice C

Scientific Notation

Write each number in standard notation.

1. 6.34×10^{5} — 634,000
2. 7.1×10^{-5} — 0.000071
3. 4.23×10^{6} — 4,230,000
4. 8.235×10^{-7} — 0.0000008235
5. 6.0089×10^{-8} — 0.000000060089
6. 5.2×10^{9} — 5,200,000,000
7. 2.0547×10^{9} — 2,054,700,000
8. 8.394×10^{-7} — 0.0000008394
9. 9.688×10^{11} — 968,800,000,000

Write each number in scientific notation.

10. 97,406,000,000 — 9.7406×10^{10}
11. 0.000000000067 — 6.7×10^{-11}
12. 5,280,000,000 — 5.28×10^{9}
13. 0.000000084 — 8.4×10^{-8}
14. 652,100,000,000 — 6.521×10^{11}
15. 0.0000000007254 — 7.254×10^{-10}
16. 20,509,000,000,000 — 2.0509×10^{13}
17. 0.000000000000066 — 6.6×10^{-14}
18. 30,500,000,000,000 — 3.05×10^{13}

19. The volume of the earth is approximately 260,000,000,000 mi^3. Write this number in scientific notation.

$2.6 \times 10^{11}\ mi^3$

20. If light travels 10,000 km in 3.3×10^{-2} sec, how long will it take light to travel one meter?

3.3×10^{-9} sec

21. The diameter of a human hair can measure from 1.7×10^{-5} meter to 1.8×10^{-4} meter. Which is the greater diameter?

1.8×10^{-4}

LESSON 4-4

Reteach

Scientific Notation

Standard Notation	Scientific Notation	
	(1st factor is between 1 and 10.)	(2nd factor is an integer power of 10.)
430,000	4.3×10^{5}	positive integer for large number
0.0000057	5.7×10^{-6}	negative integer for small number

To convert from scientific notation, look at the power of 10 to tell how many places and which way to move the decimal point.

Complete to write each in standard notation.

	1. 4.12×10^{6}	2. 3.4×10^{-5}
Is the exponent positive or negative?	positive	negative
Move the decimal point right or left? How many places?	right 6	left 5
Write the number in standard notation.	4,120,000	0.000034

Write each number in standard notation.

3. 8×10^{5} — 800,000
4. 7.1×10^{-4} — 0.00071
5. 3.14×10^{8} — 314,000,000

To convert to scientific notation, determine the factor between 1 and 10. Then determine the power of 10 by counting from the decimal point in the first factor to the decimal point in the given number.

Complete to write each in scientific notation.

	6. 32,000,000	7. 0.0000000712
What is the first factor?	3.2	7.12
From its location in the first factor, which way must the decimal move to its location in the given number? How many places?	right 7	left 8
Write the number in scientific notation.	3.2×10^{7}	7.12×10^{-8}

Write each number in scientific notation.

8. 41,000,000 — 4.1×10^{7}
9. 0.0000000643 — 6.43×10^{-8}
10. 1,370,000,000 — 1.37×10^{9}

LESSON 4-4

Challenge

The Wild Blue Yonder

Astronomers measure distances within our solar system in *astronomical units* (AU).

1 AU ≈ 92,956,000 mi or 149,600,000 km
(the distance from Earth to the Sun)

Mean Distance From the Sun

Planet	km	Scientific Notation	AU
Mercury	57,900,000	5.79×10^7	0.4
Venus	108,200,000	1.082×10^8	0.7
Earth	149,600,000	1.496×10^8	1.0
Mars	227,900,000	2.279×10^8	1.5
Jupiter	778,400,000	7.784×10^8	5.2
Saturn	1,429,400,000	1.4294×10^9	9.6
Uranus	2,875,000,000	2.875×10^9	19.2
Neptune	4,504,300,000	4.5043×10^9	30.1
Pluto*	5,900,100,000	5.9001×10^9	39.4

* designated as a dwarf planet in 2006

16

1. The table gives each planet's mean distance from the Sun in kilometers. Write these distances in scientific notation.
2. Convert to AUs by dividing each planet's mean distance from the Sun by 1.496×10^8. Use scientific notation. Round your answers to the nearest tenth of an AU.

Example $\frac{5.79 \times 10^7}{1.496 \times 10^8} = \frac{5.79}{1.496} \times \frac{10^7}{10^8} = 3.87 \times 10^{-1} = 0.387 \approx 0.4$ AU

3. Approximately how many times greater is Saturn's distance from the Sun than is Earth's? Answer to the nearest tenth.

   about 9.6 times

4. Approximately how many times greater is the distance of the farthest planet from the Sun than is the distance of the closest planet to the Sun? Answer to the nearest tenth.

   Pluto's distance from the Sun is about 101.9 times that of Mercury.

LESSON 4-4

Problem Solving

Scientific Notation

Write the correct answer.

1. In June 2001, the Intel Corporation announced that they could produce a silicon transistor that could switch on and off 1.5 trillion times a second. Express the speed of the transistor in scientific notation.

   1.5×10^{12}

2. With this transistor, computers will be able to do 1×10^9 calculations in the time it takes to blink your eye. Express the number of calculations using standard notation.

   1,000,000,000

3. The elements in this fast transistor are 20 nanometers long. A nanometer is one-billionth of a meter. Express the length of an element in the transistor in meters using scientific notation.

   2×10^{-8} m

4. The length of the elements in the transistor can also be compared to the width of a human hair. The length of an element is 2×10^{-3} times smaller than the width of a human hair. Express 2×10^{-3} in standard notation.

   0.002

Use the table to answer Exercises 5–9. Choose the best answer.

Distance From Earth To Stars
Light-Year = 5,880,000,000,000 mi.

Star	Constellation	Distance (light-years)
Sirius	Canis Major	8
Canopus	Carina	650
Alpha Centauri	Centaurus	4
Vega	Lyra	23

5. Express a light-year in miles using scientific notation.
   A 58.8×10^{11} C 588×10^{10}
   Ⓑ 5.88×10^{12} D 5.88×10^{-13}

6. How many miles is it from Earth to the star Sirius?
   F 4.705×10^{12} H 7.35×10^{12}
   Ⓖ 4.704×10^{13} J 7.35×10^{11}

7. How many miles is it from Earth to the star Canopus?
   Ⓐ 3.822×10^{15} C 3.822×10^{14}
   B 1.230×10^{15} D 1.230×10^{14}

8. How many miles is it from Earth to the star Alpha Centauri?
   Ⓕ 2.352×10^{13} H 2.352×10^{14}
   G 5.92×10^{13} J 5.92×10^{14}

9. How many miles is it from Earth to the star Vega?
   A 6.11×10^{13} C 6.11×10^{14}
   B 1.3524×10^{13} Ⓓ 1.3524×10^{14}

LESSON 4-4

Reading Strategies

Organization Patterns

You can use **powers of 10** to write very large or very small numbers in a shortened form. This efficient method is called **scientific notation.** It is also useful in performing multiplication and division of very large and very small numbers.

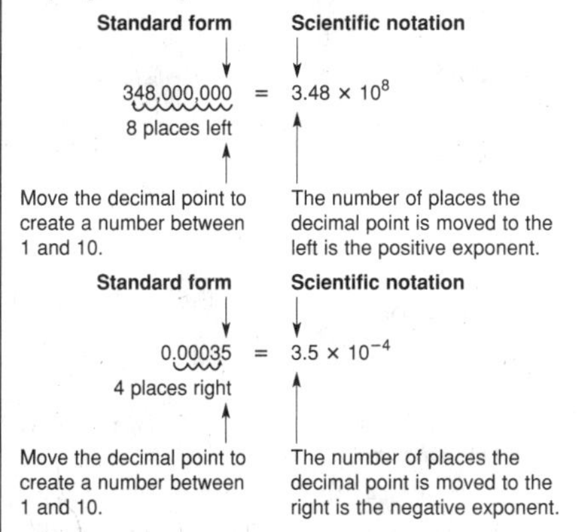

Standard form
Scientific notation
348,000,000 = 3.48 × 10^8
8 places left
Move the decimal point to create a number between 1 and 10.
The number of places the decimal point is moved to the left is the positive exponent.
Standard form
Scientific notation
0.00035 = 3.5 × 10^−4
4 places right
Move the decimal point to create a number between 1 and 10.
The number of places the decimal point is moved to the right is the negative exponent.

Use 0.000078 to answer Exercises 1–4.

1. How many places must you move the decimal point to create a number between 1 and 10? 5 places
2. Which direction will you move the decimal point? to the right
3. Will the exponent be negative or positive? negative
4. Write the number in scientific notation. 7.8×10^{-5}

Use 312,000,000 to answer Exercises 5–7.

5. How many places must you move the decimal point to create a number between 1 and 10? 8 places
6. Which direction will you move the decimal point? to the left

/6

LESSON 4-4

Puzzles, Twisters & Teasers

Be a Math Hog!

Write the correct exponents to show each number in scientific notation. Then solve the riddle.

I	$580,000 = 5.8 \times 10^{5}$
E	$26,400,000 = 2.64 \times 10^{7}$
N	$0.000135 = 1.35 \times 10^{-4}$
A	$0.000002 = 2 \times 10^{-6}$
C	$155,000,000 = 1.55 \times 10^{8}$
H	$0.00000014 = 1.4 \times 10^{-7}$
N	$0.0003 = 3 \times 10^{-4}$
M	$7,800,000 = 7.8 \times 10^{6}$
L	$0.0467 = 4.67 \times 10^{-2}$
B	$3500 = 3.5 \times 10^{3}$
U	$900 = 9 \times 10^{2}$
X	$1,000,000,000 = 1 \times 10^{9}$

How did the pig get to the hospital?

I	N		A
5	−4		−6

H	A	M	B	U	L	A	N	C	E
−7	−6	6	3	2	−2	−6	−4	8	7

/13

LESSON 4-5

Practice A

Squares and Square Roots

Find the two square roots of each number.

1. 16 **4, −4**
2. 49 **7, −7**
3. 1 **1, −1**
4. 25 **5, −5**
5. 100 **10, −10**
6. 4 **2, −2**
7. 81 **9, −9**
8. 64 **8, −8**

Evaluate each expression.

9. $\sqrt{8+1}$ **3**
10. $\sqrt{7-6}$ **1**
11. $\sqrt{18-2}$ **4**
12. $\sqrt{31+5}$ **6**
13. $\sqrt{36}+10$ **16**
14. $15-\sqrt{25}$ **10**
15. $\sqrt{49}-\sqrt{4}$ **5**
16. $\sqrt{16}+9$ **13**
17. $\sqrt{\frac{64}{16}}$ **2**
18. $5\sqrt{9}$ **15**
19. $\sqrt{\frac{100}{4}}$ **5**
20. $-3\sqrt{81}$ **−27**

Switzerland's flag is a square, unlike other flags that are rectangular.

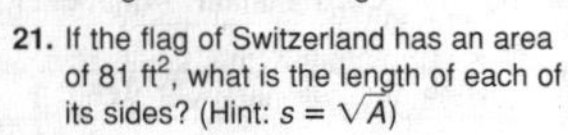

21. If the flag of Switzerland has an area of 81 ft^2, what is the length of each of its sides? (Hint: $s=\sqrt{A}$)

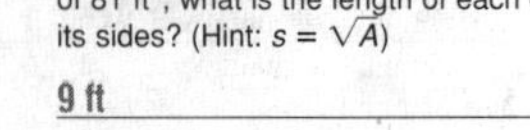

9 ft

22. If the lengths of the sides of a Switzerland flag are 10 ft, what is the area of the flag? (Hint: $A=s^2$)

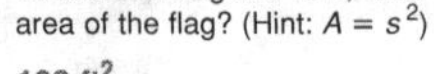

100 ft^2

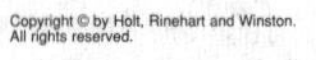

LESSON 4-5

Practice B

Squares and Square Roots

Find the two square roots of each number.

1. 36 **6, −6**
2. 81 **9, −9**
3. 49 **7, −7**
4. 100 **10, −10**
5. 64 **8, −8**
6. 121 **11, −11**
7. 25 **5, −5**
8. 144 **12, −12**

Evaluate each expression.

9. $\sqrt{32+17}$ **7**
10. $\sqrt{100-19}$ **9**
11. $\sqrt{64+36}$ **10**
12. $\sqrt{73-48}$ **5**
13. $2\sqrt{64}+10$ **26**
14. $36-\sqrt{36}$ **30**
15. $\sqrt{100}-\sqrt{25}$ **5**
16. $\sqrt{121}+16$ **27**
17. $\sqrt{\frac{25}{4}+\frac{1}{2}}$ **3**
18. $\sqrt{\frac{100}{25}}$ **2**
19. $\sqrt{\frac{196}{49}}$ **2**
20. $3(\sqrt{144}-6)$ **18**

The Pyramids of Egypt are often called the first wonder of the world. This group of pyramids consists of Menkaura, Khufu, and Khafra. The largest of these is Khufu, sometimes called Cheops. During this time in history, each monarch had his own pyramid built to bury his mummified body. Cheops was a king of Egypt in the early 26th century B.C. His pyramid's original height is estimated to have been 482 ft. It is now approximately 450 ft. The estimated completion date of this structure was 2660 B.C.

21. If the area of the base of Cheops' pyramid is 570,025 ft^2, what is the length of one of the sides of the ancient structure? (Hint: $s=\sqrt{A}$)

755 ft

22. If a replica of the pyramid were built with a base area of 625 in^2, what would be the length of each side? (Hint: $s=\sqrt{A}$)

25 in.

LESSON 4-5

Practice C

Squares and Square Roots

Find the two square roots of each number.

1. 225 **15, −15**
2. 576 **24, −24**
3. 361 **19, −19**
4. 625 **25, −25**
5. 400 **20, −20**
6. 729 **27, −27**
7. 1024 **32, −32**
8. 2500 **50, −50**

Evaluate each expression.

9. $\sqrt{\frac{1}{9}}$ **$\frac{1}{3}$**
10. $\sqrt{\frac{144}{4}}$ **6**
11. $\sqrt{\frac{64}{16}}+3$ **5**
12. $\sqrt{\frac{100}{4}}+\sqrt{4}$ **7**
13. $\sqrt{\frac{81}{324}}$

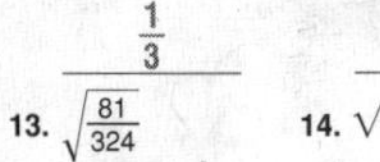

$\frac{3}{4}$

14. $\sqrt{196}-\sqrt{49}$ **7**
15. $6(\sqrt{225}-9)$ **36**
16. $-(\sqrt{144}\sqrt{64})$ **−96**

Evaluate each expression.

17. $-2\sqrt{\frac{144}{36}}-4$ **−8**

18. $\sqrt{676}-15$ **11**
19. $49-\sqrt{441}+\sqrt{169}$ **41**

20. The distance from base to base, home plate included, on a baseball field is 90 ft. The bases form a square. What is its area?

8100 ft^2

Switzerland's flag is a square, unlike other flags that are rectangular.

21. Suppose a seamstress is making a flag that has to have an area less than 50 square meters. What is the longest length, to the nearest tenth, the sides can be? (Hint: $s=\sqrt{A}$)

7.0 m

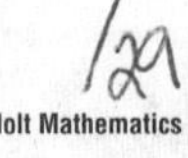

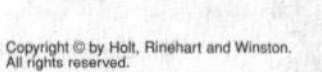

LESSON 4-5

Reteach

Squares and Square Roots

A **perfect square** has two identical factors.

$25 = 5 \times 5 = 5^2$ **or** $25 = (-5) \times (-5) = (-5)^2$ then 25 is a perfect square.

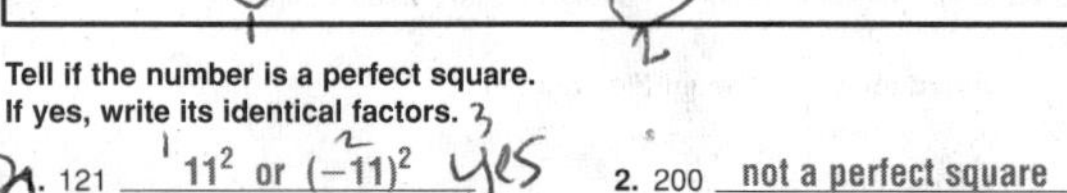

Tell if the number is a perfect square.
If yes, write its identical factors.

1. 121 **11^2 or $(-11)^2$**
2. 200 **not a perfect square**
3. 400 **20^2 or $(-20)^2$**

Since $5^2 = 25$ and also $(-5)^2 = 25$, both 5 and −5 are **square roots** of 25. $\sqrt{25} = 5$ and $-\sqrt{25} = -5$

The **principal square root** of 25 is 5: $\sqrt{25} = 5$

Write the two square roots of each number.

4. $\sqrt{81}$ = **9** $-\sqrt{81}$ = **−9**
5. $\sqrt{625}$ = **25** $-\sqrt{625}$ = **−25**
6. $\sqrt{169}$ = **13** $-\sqrt{169}$ = **−13**

Write the principal square root of each number.

7. $\sqrt{144}$ = **12**
8. $\sqrt{6400}$ = **80**
9. $\sqrt{10{,}000}$ = **100**

Use the principal square root when evaluating an expression. For the order of operations, do square root first, as you would an exponent.

$5\sqrt{100} - 3$
$5(10) - 3$
$50 - 3$
47

Complete to evaluate each expression.

10. $3\sqrt{144} - 20$
$3 \times$ **12** $- 20$
36 $- 20$
16

11. $\sqrt{25 + 144} + 13$
$\sqrt{169}$ $+ 13$
13 $+ 13$
26

12. $\sqrt{\frac{100}{25}} + \frac{1}{2}$
$\frac{\sqrt{100}}{\sqrt{25}} + \frac{1}{2}$
$\frac{10}{5} + \frac{1}{2}$
2 $+ \frac{1}{2}$
$2\frac{1}{2}$

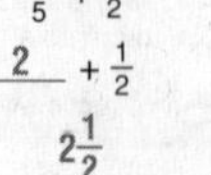

LESSON 4-5
Challenge
Dig It!

Find the **digital root** of a number by adding its digits, adding the digits of the result, and so on, until the result is a single digit.

358 → 3 + 5 + 8 = 16 → 1 + 6 = 7   The digital root of 358 is 7.

1. Complete the table to find the digital roots of the squares of 1–17.

Number	Square	Digital Root: Calculation		
1	1		=	1
2	4		=	4
3	9		=	9
4	16	1 + 6	=	7
5	25	2 + 5	=	7
6	36	3 + 6 = 9	=	9
7	49	4 + 9 = 13 → 1 + 3	=	4
8	64	6 + 4 = 10 → 1 + 0	=	1
9	81	8 + 1	=	9
10	100	1 + 0 + 0	=	1
11	121	1 + 2 + 1	=	4
12	144	1 + 4 + 4	=	9
13	169	1 + 6 + 9 = 16 → 1 + 6	=	7
14	196	1 + 9 + 6 = 16 → 1 + 6	=	7
15	225	2 + 2 + 5	=	9
16	256	2 + 5 + 6 = 13 → 1 + 3	=	4
17	289	2 + 8 + 9 = 19 → 1 + 9 = 10 → 1 + 0	=	1

2. Make an observation about the results. **Possible answers:**
   The only results are 1, 4, 7, or 9.

3. Make a conjecture about the digital root of any whole-number perfect square. Verify your conjecture by using at least three more perfect squares.
   The result is one of the numbers 1, 4, 7, or 9. Choices vary.

4. A **palindrome** is a number that is the same when read forward or backward, such as 14741. Find two palindromes in the table.
   The digital roots of the squares of the numbers 1–8 and then 10–17.

LESSON 4-5
Problem Solving
Squares and Square Roots

Write the correct answer.

1. For college wrestling competitions, the NCAA requires that the wrestling mat be a square with an area of 1764 square feet. What is the length of each side of the wrestling mat?
   42 feet

2. For high school wrestling competitions, the wrestling mat must be a square with an area of 1444 square feet. What is the length of each side of the wrestling mat?
   38 feet

3. The Japanese art of origami requires folding square pieces of paper. Elena begins with a large sheet of square paper that is 169 square inches. How many squares can she cut out of the paper that are 4 inches on each side?
   9 squares

4. When the James family moved into a new house they had a square area rug that was 132 square feet. In their new house, there are three bedrooms. Bedroom one is 11 feet by 11 feet. Bedroom two is 10 feet by 12 feet and bedroom three is 13 feet by 13 feet. In which bedroom will the rug fit?
   Bedroom three

Choose the letter for the best answer.

5. A square picture frame measures 36 inches on each side. The actual wood trim is 2 inches wide. The photograph in the frame is surrounded by a bronze mat that measures 5 inches. What is the maximum area of the photograph?
   A 841 sq. inches **B** 900 sq. inches
   C 1156 sq. inches **(D)** 484 sq. inches

6. To create a square patchwork quilt wall hanging, square pieces of material are sewn together to form a larger square. Which number of smaller squares can be used to create a square patchwork quilt wall hanging?
   F 35 squares **(G)** 64 squares
   H 84 squares **J** 125 squares

7. A can of paint claims that one can will cover 400 square feet. If you painted a square with the can of paint, how long would it be on each side?
   A 200 feet **B** 65 feet
   C 25 feet **(D)** 20 feet

8. A box of tile contains 12 square tiles. If you tile the largest possible square area using whole tiles, how many tiles will you have left from the box?
   F 9 **G** 6
   (H) 3 **J** 0

LESSON 4-5
Reading Strategies
Connect Words with Symbols

A **square root** produces a given number when multiplied by itself. The large square shown below is 4 squares long on each side and has 16 squares. 4 times 4 equals 16. 4 is the **square root** of 16.

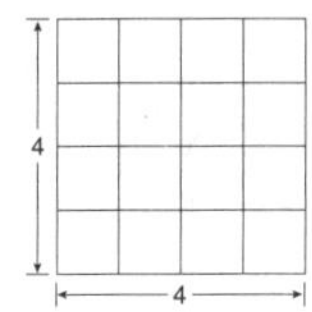

The 4 × 4 square can be described with symbols and with words.

Symbols	Symbols		Words
$4 \cdot 4 = 16$	$4^2 = 16$	→	Four squared equals sixteen.

This sign represents square root: $\sqrt{\ }$
$\sqrt{16} = 4$ → Read "The square root of 16 equals 4."
$\sqrt{25} = 5$ → Read "The square root of 25 equals 5."

Compare the symbols for "squared" and "square root."
$4^2 = 16$ and $\sqrt{16} = 4$
$5^2 = 25$ and $\sqrt{25} = 5$

Write in words.

1. 6^2 six squared
2. $\sqrt{36}$ the square root of thirty-six

Answer each question.

3. What is the square root of 36? 6
4. What is the square root of 100? 10
5. What is 7^2? 49

LESSON 4-5
Puzzles, Twisters & Teasers
Squaresville, Man!

Circle words from the list in the word search. Then find a word that answers the riddle. Circle it and write it on the line.

principal	square	root	perfect	negative
positive	solution	calculator	inverse	operation

N	M	J	P	R	I	N	C	I	P	A	L	N
E	G	H	J	O	G	B	R	L	E	J	P	O
G	M	N	B	O	S	D	U	S	R	L	O	I
A	L	K	J	T	A	X	M	E	F	O	S	T
T	C	V	B	N	M	J	B	D	E	M	I	A
I	N	V	E	R	S	E	Y	V	C	Z	T	R
V	S	Q	U	A	R	E	L	P	T	C	I	E
E	C	A	L	C	U	L	A	T	O	R	V	P
S	O	L	U	T	I	O	N	U	I	O	E	O

Why did the cookie go to see the doctor?

He was feeling C R U M B Y.

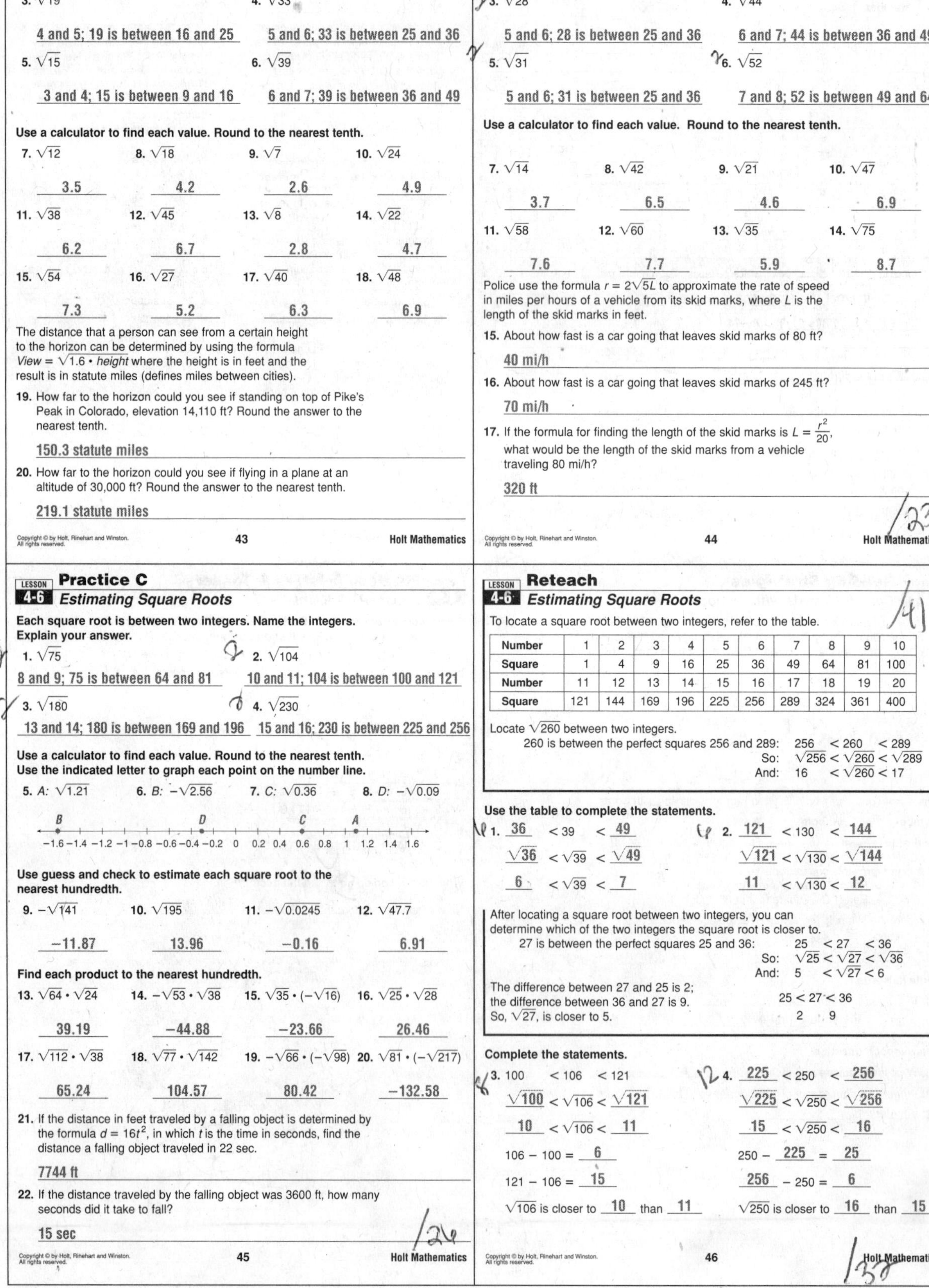

LESSON 4-6

Practice A

Estimating Square Roots

Each square root is between two integers. Name the integers. Explain your answer.

1. $\sqrt{10}$ — 3 and 4; 10 is between 9 and 16
2. $\sqrt{8}$ — 2 and 3; 8 is between 4 and 9
3. $\sqrt{19}$ — 4 and 5; 19 is between 16 and 25
4. $\sqrt{33}$ — 5 and 6; 33 is between 25 and 36
5. $\sqrt{15}$ — 3 and 4; 15 is between 9 and 16
6. $\sqrt{39}$ — 6 and 7; 39 is between 36 and 49

Use a calculator to find each value. Round to the nearest tenth.

7. $\sqrt{12}$ 3.5
8. $\sqrt{18}$ 4.2
9. $\sqrt{7}$ 2.6
10. $\sqrt{24}$ 4.9
11. $\sqrt{38}$ 6.2
12. $\sqrt{45}$ 6.7
13. $\sqrt{8}$ 2.8
14. $\sqrt{22}$ 4.7
15. $\sqrt{54}$ 7.3
16. $\sqrt{27}$ 5.2
17. $\sqrt{40}$ 6.3
18. $\sqrt{48}$ 6.9

The distance that a person can see from a certain height to the horizon can be determined by using the formula *View* = $\sqrt{1.6 \cdot \textit{height}}$ where the height is in feet and the result is in statute miles (defines miles between cities).

19. How far to the horizon could you see if standing on top of Pike's Peak in Colorado, elevation 14,110 ft? Round the answer to the nearest tenth.

    150.3 statute miles

20. How far to the horizon could you see if flying in a plane at an altitude of 30,000 ft? Round the answer to the nearest tenth.

    219.1 statute miles

LESSON 4-6

Practice B

Estimating Square Roots

Each square root is between two integers. Name the integers. Explain your answer.

1. $\sqrt{6}$ — 2 and 3; 6 is between 4 and 9
2. $\sqrt{20}$ — 4 and 5; 20 is between 16 and 25
3. $\sqrt{28}$ — 5 and 6; 28 is between 25 and 36
4. $\sqrt{44}$ — 6 and 7; 44 is between 36 and 49
5. $\sqrt{31}$ — 5 and 6; 31 is between 25 and 36
6. $\sqrt{52}$ — 7 and 8; 52 is between 49 and 64

Use a calculator to find each value. Round to the nearest tenth.

7. $\sqrt{14}$ 3.7
8. $\sqrt{42}$ 6.5
9. $\sqrt{21}$ 4.6
10. $\sqrt{47}$ 6.9
11. $\sqrt{58}$ 7.6
12. $\sqrt{60}$ 7.7
13. $\sqrt{35}$ 5.9
14. $\sqrt{75}$ 8.7

Police use the formula $r = 2\sqrt{5L}$ to approximate the rate of speed in miles per hours of a vehicle from its skid marks, where L is the length of the skid marks in feet.

15. About how fast is a car going that leaves skid marks of 80 ft?

    40 mi/h

16. About how fast is a car going that leaves skid marks of 245 ft?

    70 mi/h

17. If the formula for finding the length of the skid marks is $L = \frac{r^2}{20}$, what would be the length of the skid marks from a vehicle traveling 80 mi/h?

    320 ft

LESSON 4-6

Practice C

Estimating Square Roots

Each square root is between two integers. Name the integers. Explain your answer.

1. $\sqrt{75}$ — 8 and 9; 75 is between 64 and 81
2. $\sqrt{104}$ — 10 and 11; 104 is between 100 and 121
3. $\sqrt{180}$ — 13 and 14; 180 is between 169 and 196
4. $\sqrt{230}$ — 15 and 16; 230 is between 225 and 256

Use a calculator to find each value. Round to the nearest tenth. Use the indicated letter to graph each point on the number line.

5. *A*: $\sqrt{1.21}$
6. *B*: $-\sqrt{2.56}$
7. *C*: $\sqrt{0.36}$
8. *D*: $-\sqrt{0.09}$

Number line: −1.6 −1.4 −1.2 −1 −0.8 −0.6 −0.4 −0.2 0 0.2 0.4 0.6 0.8 1 1.2 1.4 1.6 (B at −1.6, D at −0.3, C at 0.6, A at 1.1)

Use guess and check to estimate each square root to the nearest hundredth.

9. $-\sqrt{141}$ −11.87
10. $\sqrt{195}$ 13.96
11. $-\sqrt{0.0245}$ −0.16
12. $\sqrt{47.7}$ 6.91

Find each product to the nearest hundredth.

13. $\sqrt{64} \cdot \sqrt{24}$ 39.19
14. $-\sqrt{53} \cdot \sqrt{38}$ −44.88
15. $\sqrt{35} \cdot (-\sqrt{16})$ −23.66
16. $\sqrt{25} \cdot \sqrt{28}$ 26.46
17. $\sqrt{112} \cdot \sqrt{38}$ 65.24
18. $\sqrt{77} \cdot \sqrt{142}$ 104.57
19. $-\sqrt{66} \cdot (-\sqrt{98})$ 80.42
20. $\sqrt{81} \cdot (-\sqrt{217})$ −132.58

21. If the distance in feet traveled by a falling object is determined by the formula $d = 16t^2$, in which t is the time in seconds, find the distance a falling object traveled in 22 sec.

    7744 ft

22. If the distance traveled by the falling object was 3600 ft, how many seconds did it take to fall?

    15 sec

LESSON 4-6

Reteach

Estimating Square Roots

To locate a square root between two integers, refer to the table.

Number	1	2	3	4	5	6	7	8	9	10
Square	1	4	9	16	25	36	49	64	81	100
Number	11	12	13	14	15	16	17	18	19	20
Square	121	144	169	196	225	256	289	324	361	400

Locate $\sqrt{260}$ between two integers.

260 is between the perfect squares 256 and 289: $256 < 260 < 289$

So: $\sqrt{256} < \sqrt{260} < \sqrt{289}$

And: $16 < \sqrt{260} < 17$

Use the table to complete the statements.

1. $36 < 39 < 49$; $\sqrt{36} < \sqrt{39} < \sqrt{49}$; $6 < \sqrt{39} < 7$
2. $121 < 130 < 144$; $\sqrt{121} < \sqrt{130} < \sqrt{144}$; $11 < \sqrt{130} < 12$

After locating a square root between two integers, you can determine which of the two integers the square root is closer to.

27 is between the perfect squares 25 and 36: $25 < 27 < 36$

So: $\sqrt{25} < \sqrt{27} < \sqrt{36}$

And: $5 < \sqrt{27} < 6$

The difference between 27 and 25 is 2; the difference between 36 and 27 is 9. So, $\sqrt{27}$, is closer to 5.

$25 < 27 < 36$ (differences: 2, 9)

Complete the statements.

3. $100 < 106 < 121$; $\sqrt{100} < \sqrt{106} < \sqrt{121}$; $10 < \sqrt{106} < 11$; $106 - 100 = 6$; $121 - 106 = 15$; $\sqrt{106}$ is closer to 10 than 11
4. $225 < 250 < 256$; $\sqrt{225} < \sqrt{250} < \sqrt{256}$; $15 < \sqrt{250} < 16$; $250 - 225 = 25$; $256 - 250 = 6$; $\sqrt{250}$ is closer to 16 than 15

LESSON 4-6

Challenge

Dig Deeper!

The **digital root** of a number is found by adding its digits, adding the digits of the result, and so on, until the result is a single digit.

$918 \rightarrow 9 + 1 + 8 = 18 \rightarrow 1 + 8 = 9$ The digital root of 918 is 9.

1. Complete the table to display numbers and their digital roots and to determine if they are divisible by 3 (remainder = 0). Make an observation about the results.

   A number is divisible by 3 if its digital root is divisible by 3.

Number	Divisible by 3?	Digital Root: Calculation			Divisible by 3?
81	yes	$8 + 1$	=	9	yes
92	no	$9 + 2 = 11 \rightarrow 1 + 1$	=	2	no
226	no	$2 + 2 + 6 = 10 \rightarrow 1 + 0$	=	1	no
315	yes	$3 + 1 + 5$	=	9	yes
659	no	$6 + 5 + 9 = 20 \rightarrow 2 + 0$	=	2	no
704	no	$7 + 0 + 4 = 11 \rightarrow 1 + 1$	=	2	no
1064	no	$1 + 0 + 6 + 4 = 11 \rightarrow 1 + 1$	=	2	no

2. Complete the table to display the products of numbers and the products of their digital roots. Make an observation about the results.

   The digital root of a product of whole numbers equals the product of the digital roots of the factors.

Product	Digital Root of Factor	Digital Root of Factor	Product of Digital Roots of Factors	Digital Root of Product
$24 \times 32 = 768$	$2 + 4 = 6$	$3 + 2 = 5$	$6 \times 5 = 30 \rightarrow 3 + 0 = 3$	$7 + 6 + 8 = 21 \rightarrow 2 + 1 = 3$
$11 \times 17 = 187$	$1 + 1 = 2$	$1 + 7 = 8$	$2 \times 8 = 16 \rightarrow 1 + 6 = 7$	$1 + 8 + 7 = 16 \rightarrow 1 + 6 = 7$
$121 \times 42 = 5082$	$1 + 2 + 1 = 4$	$4 + 2 = 6$	$4 \times 6 = 24 \rightarrow 2 + 4 = 6$	$5 + 0 + 8 + 2 = 15 \rightarrow 1 + 5 = 6$
$243 \times 35 = 8505$	$2 + 4 + 3 = 9$	$3 + 5 = 8$	$9 \times 8 = 72 \rightarrow 7 + 2 = 9$	$8 + 5 + 0 + 5 = 18 \rightarrow 1 + 8 = 9$
$81 \times 72 = 5832$	$8 + 1 = 9$	$7 + 2 = 9$	$9 \times 9 = 81 \rightarrow 8 + 1 = 9$	$5 + 8 + 3 + 2 = 18 \rightarrow 1 + 8 = 9$
$360 \times 54 = 19{,}440$	$3 + 6 + 0 = 9$	$5 + 4 = 9$	$9 \times 9 = 81 \rightarrow 8 + 1 = 9$	$1 + 9 + 4 + 4 + 0 = 18 \rightarrow 1 + 8 = 9$

LESSON 4-6

Problem Solving

Estimating Square Roots

The distance to the horizon can be found using the formula $d = 112.88\sqrt{h}$ where d is the distance in kilometers and h is the number of kilometers from the ground. Round your answer to the nearest kilometer.

1. How far is it to the horizon when you are standing on the top of Mt. Everest, a height of 8.85 km?

   336 km

2. Find the distance to the horizon from the top of Mt. McKinley, Alaska, a height of 6.194 km.

   281 km

3. How far is it to the horizon if you are standing on the ground and your eyes are 2 m above the ground?

   5 km

4. Mauna Kea is an extinct volcano on Hawaii that is about 4 km tall. You should be able to see the top of Mauna Kea when you are how far away?

   at most 226 km

You can find the approximate speed of a vehicle that leaves skid marks before it stops. The formulas $S = 5.5\sqrt{0.7L}$ and $S = 5.5\sqrt{0.8L}$, where S is the speed in miles per hour and L is the length of the skid marks in feet, will give the minimum and maximum speeds that the vehicle was traveling before the brakes were applied. Round to the nearest mile per hour.

5. A vehicle leaves a skid mark of 40 feet before stopping. What was the approximate speed of the vehicle before it stopped?

   A 25–35 mph
   (C) 29–31 mph
   B 28–32 mph
   D 68–70 mph

6. A vehicle leaves a skid mark of 100 feet before stopping. What was the approximate speed of the vehicle before it stopped?

   (F) 46–49 mph
   H 62–64 mph
   G 50–55 mph
   J 70–73 mph

7. A vehicle leaves a skid mark of 150 feet before stopping. What was the approximate speed of the vehicle before it stopped?

   A 50–55 mph
   C 55–70 mph
   B 53–58 mph
   (D) 56–60 mph

8. A vehicle leaves a skid mark of 200 feet before stopping. What was the approximate speed of the vehicle before it stopped?

   F 60–63 mph
   (G) 65–70 mph
   H 72–78 mph
   J 80–90 mph

LESSON 4-6

Reading Strategies

Follow a Procedure

The numbers 16 and 25 are called **perfect squares.** Each has an integer as its square root. To find the square root of a perfect square, ask yourself what number multiplied by itself equals the perfect square.

Some Perfect Squares				
1	4	9	16	25
36	49	64	81	
100	121	144	169	

1. What number times itself equals 16? 4
2. What is the square root of 16? 4
3. What number times itself equals 25? 5
4. What is the square root of 25? 5

Use these steps to estimate the square root of a number that is not a perfect square.

What is $\sqrt{45}$?

Step 1
Identify a perfect square that is a little more than $\sqrt{45}$. → $\sqrt{49}$
The square root of 49 = 7.

Step 2
Identify a perfect square that is a little less than $\sqrt{45}$. → $\sqrt{36}$
The square root of 36 = 6.

Step 3
The estimate of $\sqrt{45}$ is between 6 and 7.

Use the steps above to help you estimate the square root of 90.

5. Which perfect square is a little more than 90? 100
6. What is the square root of 100? 10
7. Which perfect square is a little less than 90? 81
8. What is the square root of 81? 9
9. What is your estimate of the square root of 90?

   between 9 and 10

LESSON 4-6

Puzzles, Twisters & Teasers

The Root of the Problem!

Find the square roots. Use the answers to solve the riddle.

S $\sqrt{36}$ = 6

P $\sqrt{100}$ = 10

G $\sqrt{25}$ = 5

W $\sqrt{4}$ = 2

E $\sqrt{81}$ = 9

L $\sqrt{121}$ = 11

R $\sqrt{144}$ = 12

T $\sqrt{64}$ = 8

I $\sqrt{9}$ = 3

H $\sqrt{169}$ = 13

U $\sqrt{49}$ = 7

Why can't you play jokes on snakes?

Because you can't P U L L (10 7 11 11) T H E I R (8 13 9 3 12) L E G S (11 9 5 6).

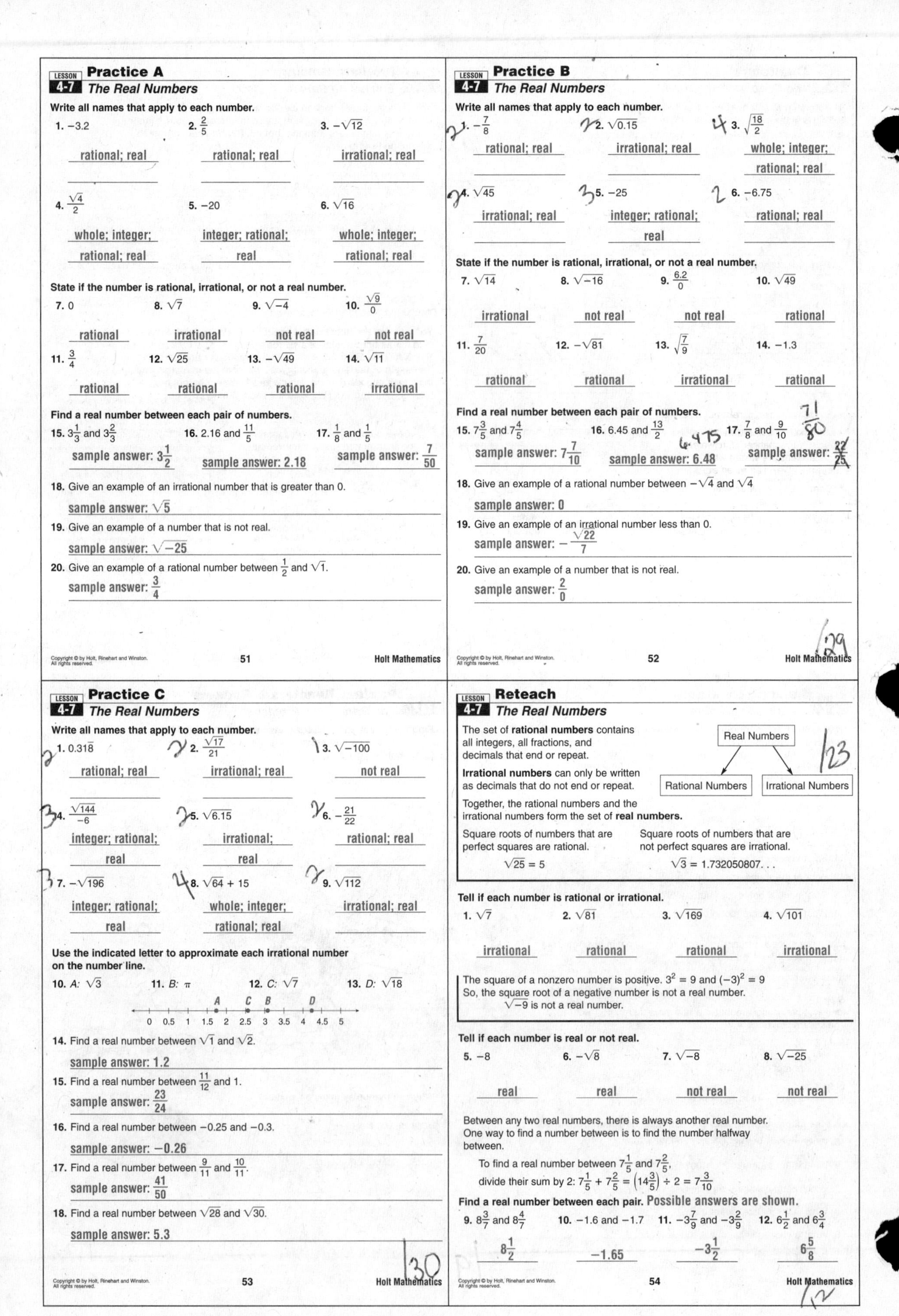

LESSON 4-7
Practice A
The Real Numbers

Write all names that apply to each number.

1. -3.2 — rational; real
2. $\frac{2}{5}$ — rational; real
3. $-\sqrt{12}$ — irrational; real
4. $\frac{\sqrt{4}}{2}$ — whole; integer; rational; real
5. -20 — integer; rational; real
6. $\sqrt{16}$ — whole; integer; rational; real

State if the number is rational, irrational, or not a real number.

7. 0 — rational
8. $\sqrt{7}$ — irrational
9. $\sqrt{-4}$ — not real
10. $\frac{\sqrt{9}}{0}$ — not real
11. $\frac{3}{4}$ — rational
12. $\sqrt{25}$ — rational
13. $-\sqrt{49}$ — rational
14. $\sqrt{11}$ — irrational

Find a real number between each pair of numbers.

15. $3\frac{1}{3}$ and $3\frac{2}{3}$ — sample answer: $3\frac{1}{2}$
16. 2.16 and $\frac{11}{5}$ — sample answer: 2.18
17. $\frac{1}{8}$ and $\frac{1}{5}$ — sample answer: $\frac{7}{50}$
18. Give an example of an irrational number that is greater than 0. — sample answer: $\sqrt{5}$
19. Give an example of a number that is not real. — sample answer: $\sqrt{-25}$
20. Give an example of a rational number between $\frac{1}{2}$ and $\sqrt{1}$. — sample answer: $\frac{3}{4}$

Copyright © by Holt, Rinehart and Winston. All rights reserved. 51 Holt Mathematics

LESSON 4-7
Practice B
The Real Numbers

Write all names that apply to each number.

1. $-\frac{7}{8}$ — rational; real
2. $\sqrt{0.15}$ — irrational; real
3. $\sqrt{\frac{18}{2}}$ — whole; integer; rational; real
4. $\sqrt{45}$ — irrational; real
5. -25 — integer; rational; real
6. -6.75 — rational; real

State if the number is rational, irrational, or not a real number.

7. $\sqrt{14}$ — irrational
8. $\sqrt{-16}$ — not real
9. $\frac{6.2}{0}$ — not real
10. $\sqrt{49}$ — rational
11. $\frac{7}{20}$ — rational
12. $-\sqrt{81}$ — rational
13. $\sqrt{\frac{7}{9}}$ — irrational
14. -1.3 — rational

Find a real number between each pair of numbers.

15. $7\frac{3}{5}$ and $7\frac{4}{5}$ — sample answer: $7\frac{7}{10}$
16. 6.45 and $\frac{13}{2}$ — sample answer: 6.48
17. $\frac{7}{8}$ and $\frac{9}{10}$ — sample answer: $\frac{22}{25}$
18. Give an example of a rational number between $-\sqrt{4}$ and $\sqrt{4}$ — sample answer: 0
19. Give an example of an irrational number less than 0. — sample answer: $-\frac{\sqrt{22}}{7}$
20. Give an example of a number that is not real. — sample answer: $\frac{2}{0}$

Copyright © by Holt, Rinehart and Winston. All rights reserved. 52 Holt Mathematics

LESSON 4-7
Practice C
The Real Numbers

Write all names that apply to each number.

1. $0.31\overline{8}$ — rational; real
2. $\frac{\sqrt{17}}{21}$ — irrational; real
3. $\sqrt{-100}$ — not real
4. $\frac{\sqrt{144}}{-6}$ — integer; rational; real
5. $\sqrt{6.15}$ — irrational; real
6. $-\frac{21}{22}$ — rational; real
7. $-\sqrt{196}$ — integer; rational; real
8. $\sqrt{64} + 15$ — whole; integer; rational; real
9. $\sqrt{112}$ — irrational; real

Use the indicated letter to approximate each irrational number on the number line.

10. A: $\sqrt{3}$
11. B: π
12. C: $\sqrt{7}$
13. D: $\sqrt{18}$

A C B D
0 0.5 1 1.5 2 2.5 3 3.5 4 4.5 5

14. Find a real number between $\sqrt{1}$ and $\sqrt{2}$. — sample answer: 1.2
15. Find a real number between $\frac{11}{12}$ and 1. — sample answer: $\frac{23}{24}$
16. Find a real number between -0.25 and -0.3. — sample answer: -0.26
17. Find a real number between $\frac{9}{11}$ and $\frac{10}{11}$. — sample answer: $\frac{41}{50}$
18. Find a real number between $\sqrt{28}$ and $\sqrt{30}$. — sample answer: 5.3

Copyright © by Holt, Rinehart and Winston. All rights reserved. 53 Holt Mathematics

LESSON 4-7
Reteach
The Real Numbers

The set of **rational numbers** contains all integers, all fractions, and decimals that end or repeat.

Irrational numbers can only be written as decimals that do not end or repeat.

Real Numbers → Rational Numbers, Irrational Numbers

Together, the rational numbers and the irrational numbers form the set of **real numbers.**

Square roots of numbers that are perfect squares are rational. $\sqrt{25} = 5$

Square roots of numbers that are not perfect squares are irrational. $\sqrt{3} = 1.732050807...$

Tell if each number is rational or irrational.

1. $\sqrt{7}$ — irrational
2. $\sqrt{81}$ — rational
3. $\sqrt{169}$ — rational
4. $\sqrt{101}$ — irrational

The square of a nonzero number is positive. $3^2 = 9$ and $(-3)^2 = 9$
So, the square root of a negative number is not a real number. $\sqrt{-9}$ is not a real number.

Tell if each number is real or not real.

5. -8 — real
6. $-\sqrt{8}$ — real
7. $\sqrt{-8}$ — not real
8. $\sqrt{-25}$ — not real

Between any two real numbers, there is always another real number. One way to find a number between is to find the number halfway between.

To find a real number between $7\frac{1}{5}$ and $7\frac{2}{5}$, divide their sum by 2: $7\frac{1}{5} + 7\frac{2}{5} = \left(14\frac{3}{5}\right) \div 2 = 7\frac{3}{10}$

Find a real number between each pair. Possible answers are shown.

9. $8\frac{3}{7}$ and $8\frac{4}{7}$ — $8\frac{1}{2}$
10. -1.6 and -1.7 — -1.65
11. $-3\frac{7}{9}$ and $-3\frac{2}{9}$ — $-3\frac{1}{2}$
12. $6\frac{1}{2}$ and $6\frac{3}{4}$ — $6\frac{5}{8}$

Copyright © by Holt, Rinehart and Winston. All rights reserved. 54 Holt Mathematics

LESSON **4-7**

Challenge

Searching for Perfection

Numbers that are equal to the sum of all their factors (not including the number itself) are called **perfect numbers.**

$6 = 1 + 2 + 3$ 6 is the smallest perfect number.

1. Which of the numbers 24 or 28 is a perfect number? Explain.

24 is not perfect since $1 + 2 + 3 + 4 + 6 + 8 + 12 \neq 24$.

28 is perfect since $1 + 2 + 4 + 7 + 14 = 28$.

The ancient Greek mathematician Euclid devised a method for computing perfect numbers.

- Begin with the number 1 and keep adding powers of 2 until you get a sum that is a *prime number* (only factors are itself and 1).
- Multiply this sum by the last power of 2.

2. Complete the table to write the first three perfect numbers.

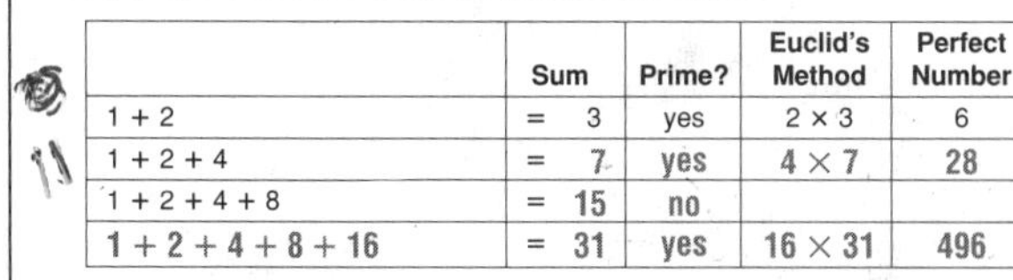

	Sum	Prime?	Euclid's Method	Perfect Number
1 + 2	= 3	yes	2 × 3	6
1 + 2 + 4	= 7	yes	4 × 7	28
1 + 2 + 4 + 8	= 15	no		
1 + 2 + 4 + 8 + 16	= 31	yes	16 × 31	496

So the first three perfect numbers are 6, 28, 496.

The next perfect number is tedious to calculate in this manner. If, however, the calculations are written with exponents, a new pattern emerges.

3. Complete the table to write the sums using exponents.

Series	Sum
$1 + 2^1$	$= 2^2 - 1$
$1 + 2^1 + 2^2$	$= 2^3 - 1$
$1 + 2^1 + 2^2 + 2^3$	$= 2^4 - 1$
$1 + 2^1 + 2^2 + 2^3 + 2^4$	$= 2^5 - 1$

Incorporating this information, Euclid proved that whenever a prime number of the form $2^n - 1$ is found, a perfect number can be written.

If $2^n - 1$ is prime, then $2^{n-1}(2^n - 1)$ is a perfect number.

4. Find the fourth perfect number. $2^6(2^7 - 1) = 8128$

5. Find the fifth perfect number. $2^{12}(2^{13} - 1) = 33{,}550{,}336$

LESSON **4-7**

Problem Solving

The Real Numbers

Write the correct answer.

1. Twin primes are prime numbers that differ by 2. Find an irrational number between twin primes 5 and 7.

Possible answer: $\sqrt{31}$

2. Rounded to the nearest ten-thousandth, $\pi = 3.1416$. Find a rational number between 3 and π.

Possible answer: $\frac{31}{10}$

3. One famous irrational number is e. Rounded to the nearest ten-thousandth $e \approx 2.7183$. Find a rational number that is between 2 and e.

Possible answer: $\frac{5}{2}$

4. Perfect numbers are those that the divisors of the number sum to the number itself. The number 6 is a perfect number because $1 + 2 + 3 = 6$. The number 28 is also a perfect number. Find an irrational number between 6 and 28.

Possible answer: $\sqrt{43}$

Choose the letter for the best answer.

5. Which is a rational number?

A the length of a side of a square with area 2 cm²
(B) the length of a side of a square with area 4 cm²
C a non-terminating decimal
D the square root of a prime number

6. Which is an irrational number?

F a number that can be expressed as a fraction
G the length of a side of a square with area 4 cm²
(H) the length of a side of a square with area 2 cm²
J the square root of a negative number

7. Which is an integer?

A the number half-way between 6 and 7
B the average rainfall for the week if it rained 0.5 in., 2.3 in., 0 in., 0 in., 0 in., 0.2 in., 0.75 in. during the week
C the money in an account if the balance was $213.00 and $21.87 was deposited
(D) the net yardage after plays that resulted in a 15 yard loss, 10 yard gain, 6 yard gain and 5 yard loss

8. Which is a whole number?

F the number half-way between 6 and 7
(G) the total amount of sugar in a recipe that calls for $\frac{1}{4}$ cup of brown sugar and $\frac{3}{4}$ cup of granulated sugar
H the money in an account if the balance was $213.00 and $21.87 was deposited
J the net yardage after plays that resulted in a 15 yard loss, 10 yard gain, 6 yard gain and 5 yard loss

LESSON **4-7**

Reading Strategies

Use a Venn Diagram

You know that **rational numbers** can be written in fraction form as an $\frac{\text{integer}}{\text{integer}}$. Rational numbers include:

- **Decimals** • **Fractions** • **Integers** • **Whole Numbers**

This diagram of rational numbers expressed in different forms helps you see how they are related.

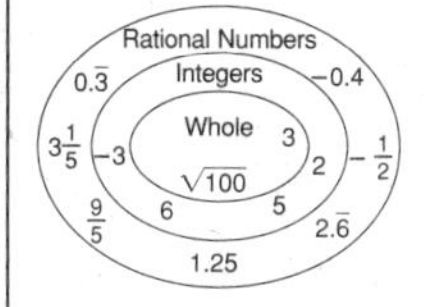

From this picture you can say:

1. −0.4 is a rational number, but it is not an integer or a whole number

2. $\sqrt{100} = 10$. It is a rational number, it is an integer, and it is a whole number.

3. −3 is a rational number and an integer, but it is not a whole number.

4. $2.\overline{6}$ is a rational number, but it is not an integer or a whole number.

Numbers that are not rational are called **irrational numbers**. For example, $\sqrt{3}$ is an irrational number. It is a decimal that does not terminate or repeat. $\sqrt{3} = 1.7320508...$

Write all names that apply to each number: rational, irrational, integer, or whole number.

5. 2.236068... irrational

6. −7 integer, rational

7. 328 rational, integer, whole number

8. $2\frac{2}{3}$ rational

LESSON **4-7**

Puzzles, Twisters & Teasers

Get Real!

Circle the correct word to complete the sentences. Then use the question numbers and answer letters as your key to unlocking the answer to the riddle.

1. ________ numbers can be written as fractions or as decimals that either terminate or repeat.

(S) Rational T Irrational

2. ________ numbers can only be written as decimals that do not terminate or repeat.

B Rational (D) Irrational

3. A repeating decimal will ________ appear to repeat on a calculator.

(K) sometimes L always

4. The set of ________ numbers consists of the set of rational numbers and the set of irrational numbers.

E whole (O) real

5. The square root of a negative number ________ a real number.

S is (C) is not

6. The density property of real numbers states that between any two real numbers is another ________ number.

M unreal (N) real

7. There ________ an integer between −2 and −3.

P is (L) is not

8. 0.25 would be an example of a ________ decimal.

S repeating (T) terminating

9. Ten is a whole number that ________ a perfect square.

O is (E) is not

10. A number and its reciprocal have a product of ________.

(P) one B zero

Why are pianos hard to open?

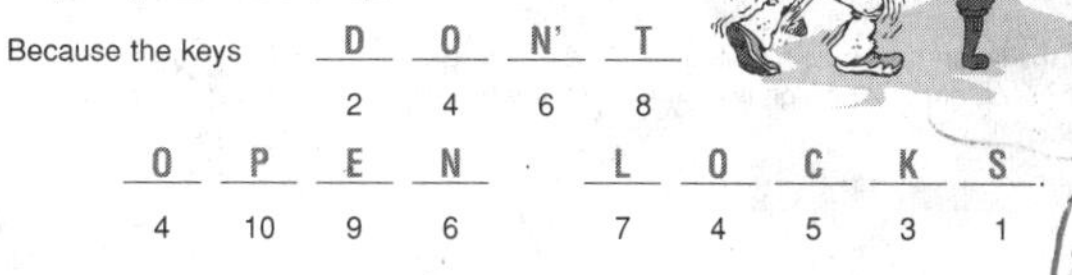

Because the keys D O N' T (2 4 6 8) O P E N (4 10 9 6) L O C K S (7 4 5 3 1)

LESSON **4-8**
Practice A
The Pythagorean Theorem

Find the length of the hypotenuse in each triangle using the Pythagorean Theorem, $a^2 + b^2 = c^2$.

1. 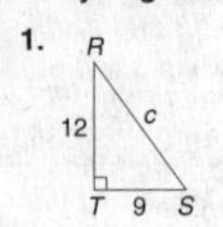

15

2. 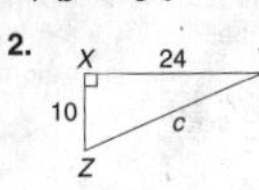

26

3. 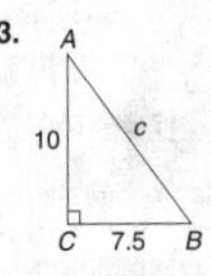

12.5

Solve for the unknown side in each right triangle. Round the answers to the nearest tenth.

4. 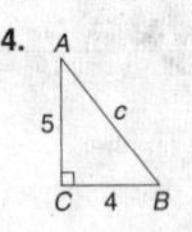

6.4

5. 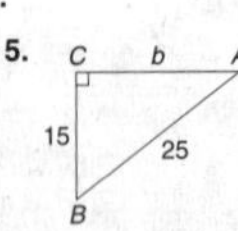

20

6. 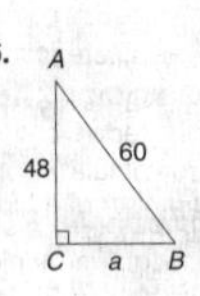

36

7. 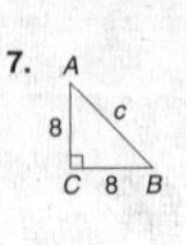

11.3

8. 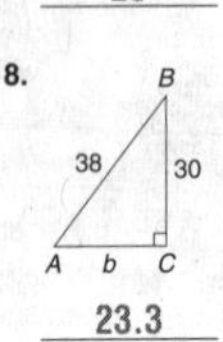

23.3

9. 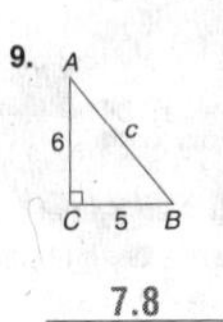

7.8

10. Jan and her brother Mel go to different schools. Jan goes 6 kilometers east from home. Mel goes 8 kilometers north. How many kilometers apart are their schools?

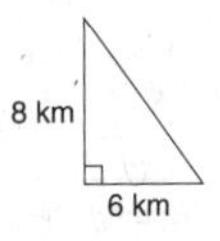

10 km

LESSON **4-8**
Practice B
The Pythagorean Theorem

Find the length of the hypotenuse to the nearest tenth.

1. 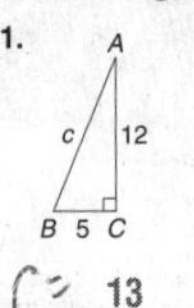

C= 13

2. 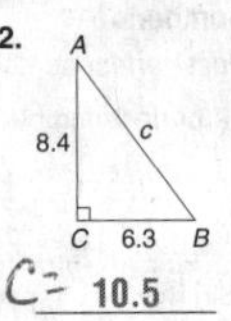

C= 10.5

3. 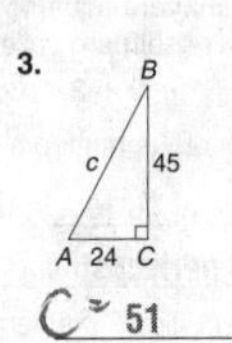

C≈ 51

Solve for the unknown side in each right triangle to the nearest tenth.

4. 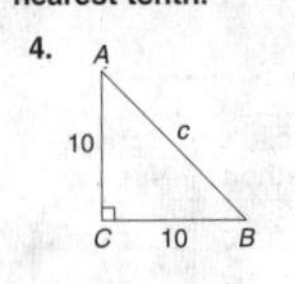

C= 14.1

5. 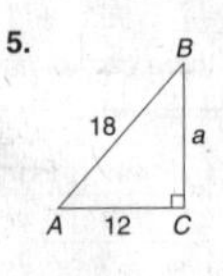

a= 13.4

6. 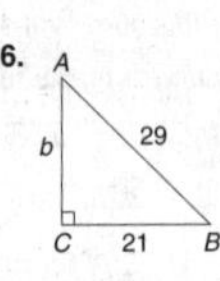

b= 20

7. A b C, 23, 14, B

b= 18.2

8. 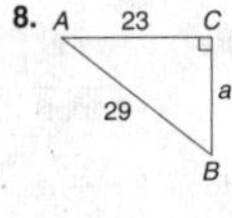

a= 17.7

9. 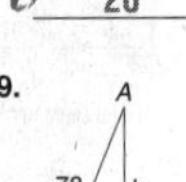 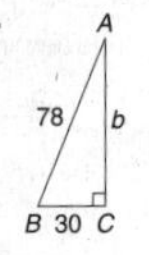

b= 72

10. A glider flies 8 miles south from the airport and then 15 miles east. Then it flies in a straight line back to the airport. What was the distance of the glider's last leg back to the airport?

17 mi

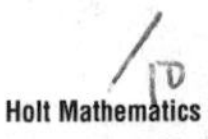

LESSON **4-8**
Practice C
The Pythagorean Theorem

Solve for the unknown side in each right triangle to the nearest tenth..

1. 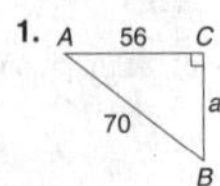

42

2. 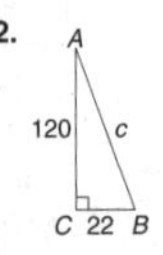

122

3. 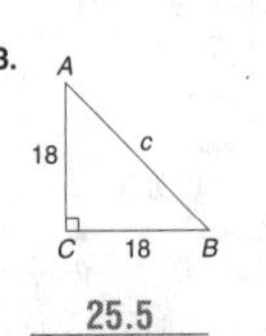

25.5

4. $a = 8$, $b = 15$, $c = ?$ — 17

5. $a = 0.5$, $b = ?$, $c = 1.3$ — 1.2

6. $a = ?$, $b = 18$, $c = 28$ — 21.4

7. $a = 21$, $b = ?$, $c = 46$ — 40.9

8. $a = ?$, $b = 38$, $c = 45$ — 24.1

9. $a = 30$, $b = ?$, $c = 50$ — 40

10. $a = 30$, $b = 72$, $c = ?$ — 78

11. $a = 40$, $b = ?$, $c = 65$ — 51.2

12. $a = 65$, $b = ?$, $c = 97$ — 72

Determine whether each set is a Pythagorean triple.

13. 2.1, 2.8, 3.5 — yes

14. 12, 15, 20 — no

15. 30, 70, 78 — no

16. 18, 24, 30 — yes

17. Use the Pythagorean Theorem to find the missing side of the triangle if the hypotenuse is 68 and the other side is 32.

60

18. Use the Pythagorean Theorem to find the base of this triangle.

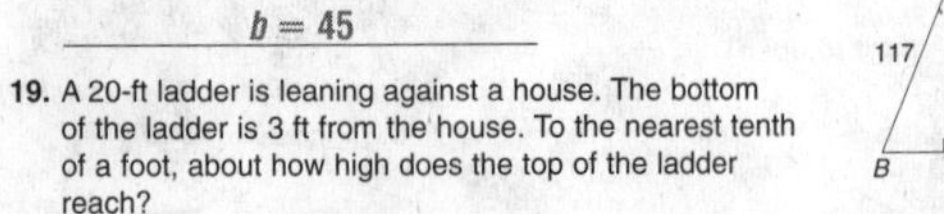

$b = 45$

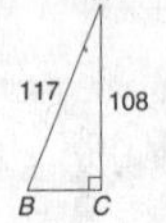

19. A 20-ft ladder is leaning against a house. The bottom of the ladder is 3 ft from the house. To the nearest tenth of a foot, about how high does the top of the ladder reach?

19.8

LESSON **4-8**
Reteach
The Pythagorean Theorem

In a right triangle,
the sum of the areas of the squares on the legs
is equal to
the area of the square on the hypotenuse.

$3^2 + 4^2 = 5^2$

$9 + 16 = 25$

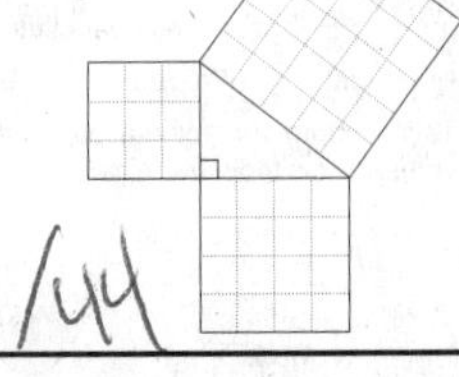

Given the squares that are on the legs of a right triangle, draw the square for the hypotenuse.

1. leg — leg — hypotenuse

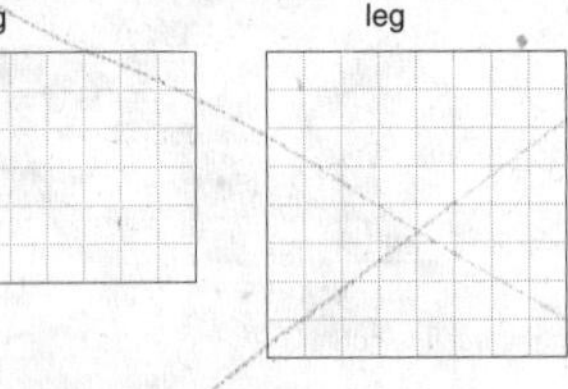

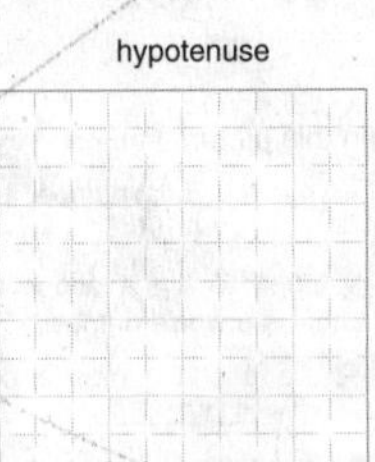

Without drawing the squares, you can find the length of a side.

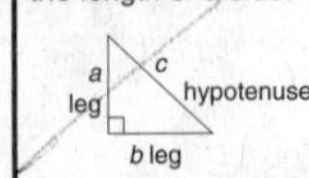

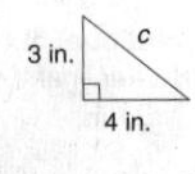

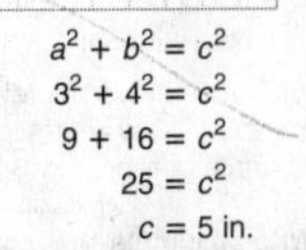

$a^2 + b^2 = c^2$
$3^2 + 4^2 = c^2$
$9 + 16 = c^2$
$25 = c^2$
$c = 5$ in.

Complete to find the length of each hypotenuse.

2. 5 ft, 12 ft, c

$a^2 + b^2 = c^2$

$5^2 + 12^2 = c^2$

$25 + 144 = c^2$

$169 = c^2$

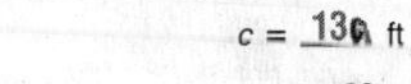

$c = 13$ ft

3. 8 in., 15 in., c

$a^2 + b^2 = c^2$

$8^2 + 15^2 = c^2$

$64 + 225 = c^2$

$289 = c^2$

$c = 17$ in.

LESSON 4-8 **Reteach**

The Pythagorean Theorem (continued)

You can use the Pythagorean Theorem to find the length of a leg if you know the length of the other leg and the hypotenuse.

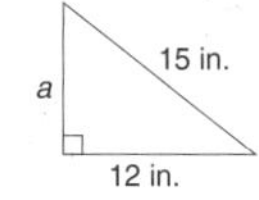

$a^2 + b^2 = c^2$

$a^2 + \underline{12^2} = \underline{15^2}$

$a^2 + \underline{144} = \underline{225}$

$-\ \underline{144} \quad -\ \underline{144}$

$a^2 = \underline{81}$

$a = \underline{9}$ in.

Complete to find the length of each leg.

4.

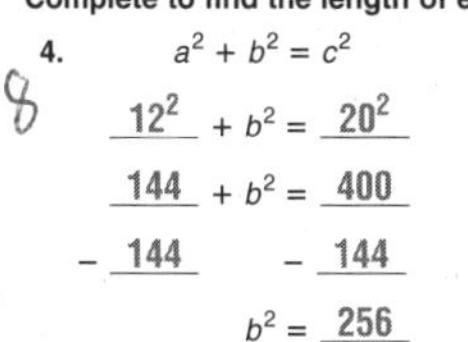

$a^2 + b^2 = c^2$

$\underline{12^2} + b^2 = \underline{20^2}$

$\underline{144} + b^2 = \underline{400}$

$-\ \underline{144} \quad -\ \underline{144}$

$b^2 = \underline{256}$

$b = \underline{16}$ in.

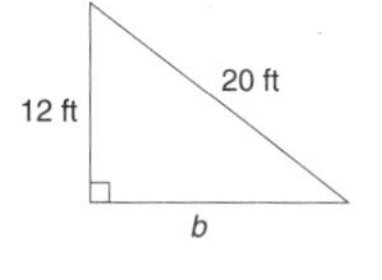

5.

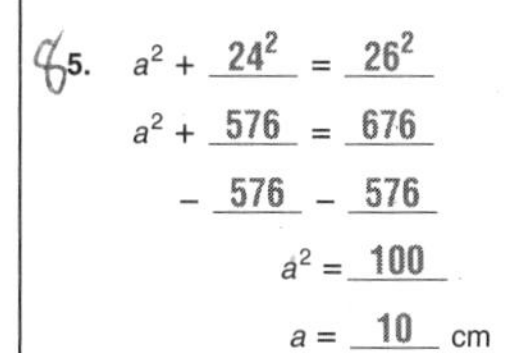

$a^2 + \underline{24^2} = \underline{26^2}$

$a^2 + \underline{576} = \underline{676}$

$-\ \underline{576} \quad -\ \underline{576}$

$a^2 = \underline{100}$

$a = \underline{10}$ cm

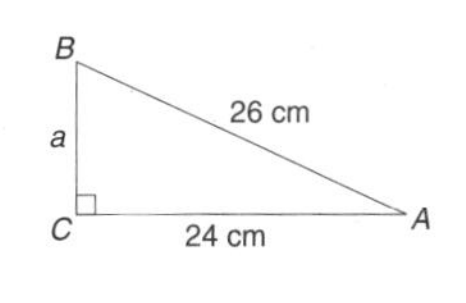

LESSON 4-8 **Challenge**

Triple Play

Three numbers connected by the Pythagorean relation are called **Pythagorean triples.**

Since $3^2 + 4^2 = 5^2$, the numbers 3-4-5 are a Pythagorean Triple.

Consider the Pythagorean triples shown in the table.

	Column *A*	Column *B*	Column *C*
row 1	3	4	5
row 2	5	12	13
row 3	7	24	25
row 4	9	40	41
row 5	11	60	61

1. Make an observation about the numbers in Column *A*.

consecutive odd numbers

2. How are the numbers in Column *C* related to those in Column *B*?

$C = B + 1$

3. Complete this table by carrying out the indicated calculation. Two calculations are done.

	Column *A*	row × *A* + row
row 1	3	1 × 3 + 1 = 4
row 2	5	2 × 5 + 2 = 12
row 3	7	3 × 7 + 3 = 24
row 4	9	4 × 9 + 4 = 40
row 5	11	5 × 11 + 5 = 60

Compare the results to the Pythagorean triples in Columns *A*, *B*, and *C* of the original table.

results = Column *B*

4. In the original table, how do the squares of the numbers in Column *A* relate to the numbers in Columns *B* and *C*?

$A^2 = B + C$

5. Using the relationships you have observed, calculate rows 6 and 10 of the table of Pythagorean triples. Verify your results by applying the Pythagorean Theorem.

	Column *A*	Column *B*	Column *C*	Verify $A^2 + B^2 = C^2$
row 6	13	84	85	$13^2 + 84^2 \stackrel{?}{=} 85^2$; $7225 = 7225$ ✓
row 10	21	220	221	$21^2 + 220^2 \stackrel{?}{=} 221^2$; $48,841 = 48,841$ ✓

LESSON 4-8 **Problem Solving**

The Pythagorean Theorem

Write the correct answer. Round to the nearest tenth.

1. A utility pole 10 m high is supported by two guy wires. Each guy wire is anchored 3 m from the base of the pole. How many meters of wire are needed for the guy wires?

20.9 m

2. A 12 foot-ladder is resting against a wall. The base of the ladder is 2.5 feet from the base of the wall. How high up the wall will the ladder reach?

11.7 ft

3. The base-path of a baseball diamond form a square. If it is 90 ft from home to first, how far does the catcher have to throw to catch someone stealing second base?

127.3 ft

4. A football field is 100 yards with 10 yards at each end for the end zones. The field is 45 yards wide. Find the length of the diagonal of the entire field, including the end zones.

128.2 yd

Choose the letter for the best answer.

5. The frame of a kite is made from two strips of wood, one 27 inches long, and one 18 inches long. What is the perimeter of the kite? Round to the nearest tenth.

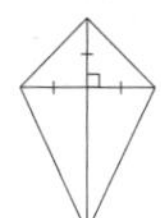

A 18.8 in.
(C) 65.7 in.
B 32.8 in.
D 131.2 in.

6. The glass for a picture window is 8 feet wide. The door it must pass through is 3 feet wide. How tall must the door be for the glass to pass through the door? Round to the nearest tenth.

F 3.3 ft
(H) 7.4 ft
G 6.7ft
J 8.5 ft

7. A television screen measures approximately 15.5 in. high and 19.5 in. wide. A television is advertised by giving the approximate length of the diagonal of its screen. How should this television be advertised?

(A) 25 in.
C 12 in.
B 21 in.
D 6 in.

8. To meet federal guidelines, a wheelchair ramp that is constructed to rise 1 foot off the ground must extend 12 feet along the ground. How long will the ramp be? Round to the nearest tenth.

F 11.9 ft
H 13.2 ft
(G) 12.0 ft
J 15.0 ft

LESSON 4-8 **Reading Strategies**

Vocabulary

Right triangles are a special type of triangle. They have one **right angle.** The right angle measures 90°. The angle may be marked with a small square.

The two shorter sides that form the right angle of a triangle are called **legs.** The third and longest side of a right triangle is called the **hypotenuse.** The hypotenuse is always located opposite the right angle.

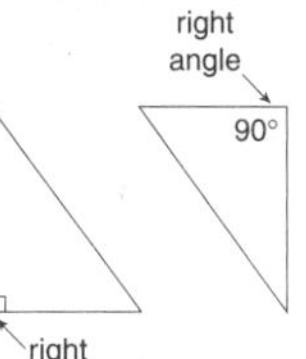

leg
hypotenuse
leg

Answer each question.

1. How many right angles does a right triangle have?

one

2. What name is given to the two sides of the triangle that form the right angle?

legs

3. What symbol is used to identify a right angle?

a small square

4. How many degrees are in a right angle?

90°

5. What name is given to the longest side of a right triangle?

the hypotenuse

6. What is the name of the side opposite the right angle?

the hypotenuse

7. Is it possible for a right triangle to have two right angles? Why or why not?

No; Possible explanation: If there were two right angles, then two sides would never meet to form a triangle.

LESSON 4-8 **Puzzles, Twisters & Teasers**

The Root of the Problem!

Find the length of the hypotenuse in each triangle. Each answer has a corresponding letter. Use the letters to solve the riddle.

1. T = 1.4

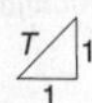

2. Y = 10.6

3. D = 13.4

4. N = 10

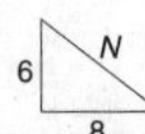

5. A = 8.6

6. I = 5.4

If a dog is tied to a rope 15 feet long, how can it reach a bone 30 feet away? The rope isn't

T I E D TO A N Y T H I N G

I	D	A	Y	T	N
5.4	13.4	8.6	10.6	1.4	10